1994

LECTURES ON
APPLICATIONS-ORIENTED
MATHEMATICS

LECTURES ON APPLICATIONS-ORIENTED MATHEMATICS

By

Bernard Friedman
Late Professor of Mathematics
University of California, Berkeley

Edited by

Victor Twersky
Professor of Mathematics
University of Illinois, Chicago

Wiley Classics Library Edition Published 1991

A Wiley-Interscience Publication
JOHN WILEY & SONS, INC.
New York / Chichester / Brisbane / Toronto / Singapore

This book is based primarily on the following reports of the
Electronic Defense Laboratories of Sylvania Electronics
Systems-West of General Telephone and Electronics, Inc.:
EDL-L-Series reports 6, 7, 8, 11, 12, 15, 18, 19, and 22.

This book was originally published in 1969 by Holden-Day, Inc.

In recognition of the importance of preserving what has been
written, it is a policy of John Wiley & Sons, Inc., to have books
of enduring value published in the United States printed on
acid-free paper, and we exert our best efforts to that end.

Library of Congress Cataloging in Publication Data:
Friedman, Bernard, 1915-1966.
　　　　　Lectures on applications-oriented mathematics / Bernard Friedman ;
　　　　　edited by Victor Twersky. --Wiley classics library ed.
　　　　　　　p.　　cm. -- (Wiley classics library)
　　　　　"A Wiley-Interscience publication."
　　　　　Includes bibliographical references (p.　　　)
　　　　　ISBN 0-471-54290-3 (paper)
　　　　　1. Mathematics. I. Twersky, Victor, 1923-　. II. Title.
　　III. Series.
　　QA37.2.F74　1991
　　510—dc20
　　　　　　　　　　　　　　　　　　　　　　　　　91-216
　　　　　　　　　　　　　　　　　　　　　　　　　CIP

Printed in the United States of America

10　9　8　7　6　5　4　3　2　1

INTRODUCTION

During the years 1958-1965 the late Professor Bernard Friedman gave a series of eleven short lecture courses in mathematics at Sylvania's Electronic Defense Laboratories in Mountain View, California. Most of the courses dealt with major topics of direct interest to applications in physics and engineering, but several provided relatively concrete introductions to topics of abstract mathematics. This book consists of the revised and edited student notes for eight of those lecture courses.

The idea of the program of courses goes back to the early 1950's when Friedman and I shared an office at the Courant Institute of Mathematical Sciences of New York University. The shortcomings of the practically antimathematical survey courses in techniques were clear. It was also clear that nonmathematics students could not afford a fuller program in mathematics at the expense of material closer to their main interests, and that they regarded many conventional mathematics courses as diversionary, with too little time allotted to potentially applicable methods. There was need for a program of courses to distill the essentials of various mathematical topics for presentation to an applications-conscious audience. The courses were to be in mathematics rather than in its applications, but some applications would be considered for illustrative purposes. Proofs were not to be stressed, but representative proofs would be included as vehicles to discuss limitations of procedures and of results; although applications-oriented, the courses would also include selected topics of abstract mathematics (as much for aesthetics as for eventual utility) taught in a more concrete way than customary. We talked about a set of short courses (ten to fifteen hours per course), some to serve as practical appendices to the conventional mathematics courses usually taken by physics and engineering students, and others to serve as introductions and provide motivation for self-study in areas of abstract mathematics.

The opportunity to develop the program came later when Friedman was Professor of Mathematics at the University of California in Berkeley and I was Head of Research at Sylvania in Mountain View. The lectures were supported by Sylvania of General Telephone and Electronics, Inc., as part of the advanced education program set up under the auspices of Jesse Lein, the Vice-President and General Manager in Mountain View. The audience consisted of twenty to forty physicists, engineers, and mathematicians; practically all had Master's degrees, and over half had Ph.D.'s. The class met weekly for a two-hour lecture at a fixed time during the five to seven successive weeks that the particular course ran. The students were provided with Friedman's *Principles and Techniques of Applied Mathematics* (Wiley, New York, 1956) as the basic reference for the program, and the major topics for the individual lecture courses were selected by polling the potential audience. (Additional publications that were cited, or that

v

are appropriate, for the individual courses are listed in the short bibliography at the end of the book.) Various attendees, particularly John Downy, Maynard Stevenson, Suzanne Lyttle, and Sandra Hawley, took detailed notes of the lectures, and those sets of notes that Friedman edited were prepared by Shirley Radl for publication in the Sylvania EDL "L-series" of reports. The eight chapters of this book correspond to his first eight lecture courses; his last three courses in the program were Wave Propagation, Numerical Methods, and Integral Equations.

Friedman had intended to edit the notes and reports fully for publication as a volume of lectures, and he had corresponded with Holden-Day on the matter in 1965. After the summer of 1966, Dorothy Friedman completed the arrangements. In helping to make Professor Friedman's material generally available, I have attempted to maintain the full flavor of his individual approach to the topics and of his unique lecture-style presentation. He was very very quick at the board, but he would always let himself be interrupted to elaborate on any aspect of the subject. His breadth and depth in mathematics, and his patience and kindness were extraordinary: he would listen to anything and everything, and would then digest and rephrase until the student saw the real root of his difficulty. He kept to a moderate, non-fussy level of rigor for an applications-minded audience, but he would shift from high-gear to low to supply details for a mathematically unsophisticated listener, and into overdrive to sketch generalizations requested by a better prepared student. The variation in pace was part of the great charm of his delivery. Various sections of this book provide almost literal records of his board-work, and are at least suggestive of his style. Some gaps in the notes have been filled with his own material from his reports and lecture notes of the Courant Institute of Mathematical Sciences of New York University, of the Mathematics Research Laboratory of the Boeing Scientific Research Laboratories, and of the Mathematics Department of the University of California, Berkeley. The work owes much to the cooperation and efforts of Dr. James E. Burke of Sylvania, Dr. Burton H. Colvin of Boeing, and Professor Harold Grad of New York University, in providing the notes and reports that were required. Professor Grad also went through various sections of the final version to help insure that the material was along the lines that Friedman had intended.

The individual lecture courses were practically independent and slight redundancies have been kept so that the following eight chapters are similarly independent. The order is chronological and thus indicates the priority of mathematical needs expressed by the Laboratories' research-minded personnel. Roughly half the material deals with techniques, and half with concepts. The techniques-heavy chapters, in an order perhaps more appropriate for self-study, are Difference Equations (4), Complex Integration (5), Distributions (1), Asymptotic Methods (3), and Perturbation Theory (8); however, the last is related to the more abstract material, particularly on Spectral Theory of Operators (2), and (2) in turn is linked to Symbolic Methods (6) in its discussion of noncommutative operators, and to Probability (7) in its treatment of event spaces. (Note that although some topics are considered in more than one chapter, e.g., the Laplace transform in Chapters 1 and 6, the discussions are independent.) Aside from the interest of the material for self-study, various chapters can also serve as supplements in the current restructuring and stretching of traditional one-semester courses to fit a two-quarter format.

Friedman's research covered a very broad spectrum ranging from fundamental work in function theory, eigenvalue problems, asymptotic integration, hydrodynamics, and electromagnetic theory, on to practical applications of mathematics in radio-wave propagation. The level of his contributions is indicated by his having served as vice-president for mathematics of the American Association of the Advancement of Science (his retiring address "What Are Mathematicians Doing" [*Science*, **154**, 357-362 (Oct. 21, 1966)] summarizes his over-all view of mathematics), and as a member of the U.S.A. Commission VI of the International Scientific Radio Union (URSI). His principal research and teaching was at New York University (1943-1957) and at the University of California in Berkeley (1957-1966), and he consulted for the Atomic Energy Commission, the Brookhaven and Argonne National Laboratories, Sylvania, Boeing, and for other governmental and industrial organizations. He was also very interested in school mathematics and teachers programs, on which he worked at the University of Illinois in Urbana during 1963-1964, and at Stanford University during several summers.

His broad ranges of interests and abilities in mathematics and science were called into full play in his lectures on applications-oriented mathematics: he structured the development of a topic to exploit the intuition and background of the physicist and engineer, and although the initial level was elementary, he used procedures novel enough so that the mathematicians in the audience were never bored. I feel that his deepest fulfillment came from his research and teaching on the interplay of mathematics and science. He would discuss physics abstractly for mathematicians, and he would motivate abstruse concepts by what physicists regard as congenital, or at most cradle,mathematics. He was a vital and enthusiastic lecturer who could simultaneously stimulate physicists to learn mathematics and mathematicians to learn physics.

Friedman's various lecture courses were received enthusiastically by all components of the audience. He was always in demand for additional courses, but his other commitments (he was also Chairman of the Mathematics Department at Berkeley for three years during this period) limited the time he had available. As the word went out that the L-reports were available, the Laboratories received scores of requests for copies. The original stocks were soon exhausted, and all the lecture reports were reprinted at least once. Although the material as it stands is a poor substitute for the vigorous lectures, or for the detailed revision that Friedman intended, the favorable reception given the original reports indicates that this collection will fill a definite need.

I am privileged to have audited Friedman's courses and to have had him as a most congenial office mate and consultant and personal teacher while at N.Y.U., and to have continued to work closely with him on a variety of research and educational programs in the years that followed. He is irreplaceable as a friend as a teacher as a scientist.

Victor Twersky
June, 1967

NOTE OF APPRECIATION

I am deeply indebted to Professor Victor Twersky for his loyal friendship and unselfish devotion to the late Professor Bernard Friedman.

Despite his own extermely busy schedule, Professor Twersky freely and voluntarily gave of his time to see that the work of Professor Friedman continues to live and that Professor Friedman continues to teach generations of students.

DOROTHY M. FRIEDMAN
(Mrs. Bernard Friedman)

TABLE OF CONTENTS

CHAPTER 1

Distributions

1.1 DELTA FUNCTIONS AND DISTRIBUTIONS

INTRODUCTION

The theory of distributions was developed by Laurent Schwartz in order to give a mathematically rigorous basis for the use of delta functions. An insight into the motivation of his theory will be obtained by our trying to understand the meaning of a charge distribution. Engineers and physicists have always used many kinds of charge distributions such as point charge, line charge, surface charge, volume charge, point dipole, surface layer of dipoles, and so on. We shall try to give a mathematically precise meaning to such terms.

Let us begin by considering the physics of the situation. How can we tell what kind of a charge distribution we have? The only way is by making an experiment which will measure the charge. We use an instrument whose reading gives us some kind of average of the charge contained in a region. The size of the region depends upon the kind of instrument we use. If we want to find the charge at a point, we must use a set of instruments that measure the charge in smaller and smaller regions containing the point of interest. We act as if we believe that if quantum effects are neglected, we can get an instrument that measures the charge in an arbitrarily small region containing the point, but both theoretically and practically, we can never get an instrument to measure the charge exactly at a point.

We may describe the situation in this way: corresponding to every instrument φ, there is a number $Q(\varphi)$, the reading of that instrument, which measures the charge. For example, if there exists a volume charge density $\rho(\vec{x})$, then $Q(\varphi)$ might be the integral $Q(\varphi) = \int \varphi(\vec{x})\rho(\vec{x})d\vec{x}$. This integral has the following obvious properties:

$$Q(\alpha\varphi) = \int \alpha\varphi(\vec{x})\rho(\vec{x})\,d\vec{x} = \alpha Q(\varphi)\,, \tag{1}$$

where α is any scalar (any number), and

$$Q(\varphi_1 + \varphi_2) = \int [\varphi_1(\vec{x}) + \varphi_2(\vec{x})]\rho(\vec{x})\,d\vec{x} = Q(\varphi_1) + Q(\varphi_2)\,. \tag{2}$$

We shall assume that our instruments always have the properties (1) and (2). Mathematically, this means that $Q(\varphi)$ is a *linear functional* on the set of instruments. We shall call the function $\varphi(\vec{x})$ which corresponds to the instrument φ a *testing function*. We shall assume that a testing function is a continuous function having continuous derivatives of all orders and that the testing function *decreases rapidly at infinity*. This last requirement means that any derivative of $\varphi(\vec{x})$ multiplied by any power of the distance goes to zero as the length of $\vec{x}$ goes to infinity. Mathematically, this requirement is convenient because it ensures convergence of the integrals we use. Physically, this requirement is reasonable because it corresponds to the fact that any instrument measures only the charge in a finite region of space. An example of a testing function is given by the function $\varphi(\vec{x}) = \exp(-|\vec{x}|^2)$.

TESTING FUNCTIONS AND DISTRIBUTIONS

For simplicity of exposition, we shall restrict ourselves to a consideration of one-dimensional distributions. In this case a testing function $\varphi(x)$ is a function which is continuous, all of whose derivatives are continuous, and for which, given any integers m and n, there exists a number M, depending on m and n, such that

$$\left| x^m \frac{d^n \varphi(x)}{dx^n} \right| < M . \tag{3}$$

Suppose that corresponding to every testing function $\varphi(x)$ there is a number Q_φ which has the following properties:

(a) If α is any scalar, then $Q_{\varphi\alpha} = \alpha Q_\varphi$.

(b) If φ_1 and φ_2 are testing functions, then $Q_{\varphi_1+\varphi_2} = Q_{\varphi_1} + Q_{\varphi_2}$.

These are exactly the properties (1) and (2) that we found before in the case where Q_α was defined by a charge density. We shall say that in all cases the linear operator Q_φ defines a *distribution*.

SYMBOLIC FUNCTIONS

When the distribution is defined by a charge density $\rho(x)$ we may write

$$Q_\varphi = \int \rho(x)\varphi(x) \, dx . \tag{4}$$

Note that all integrals will be from $-\infty$ to $+\infty$ unless specific limits are indicated. For some distributions there may not exist a charge density, and we cannot write Q_φ in the form of an integral. For example, consider the distribution defined as follows: the measurement given by an instrument φ is just the value of $\varphi(x)$ at $x = 0$; that is, $Q_\varphi = \varphi(0)$. It is easy to prove that there does not exist a function $\rho(x)$ such that $\int \rho(x)\varphi(x) \, dx = \varphi(0)$ for every testing function $\varphi(x)$. Nevertheless, the integral representation is so convenient and so useful that we shall introduce a *symbolic function* $\delta(x)$ such that

$$\int \delta(x)\varphi(x) \, dx = \varphi(0) . \tag{5}$$

The symbol $\delta(x)$ is not a function of x: it is meaningless to ask for the value of $\delta(x)$ when $x = 5$, say. The only expression containing $\delta(x)$ to which we assign a number is the integral $\int \delta(x)\varphi(x) \, dx$, where $\varphi(x)$ is any testing function.

We shall do a similar thing for certain more general distributions. Given a distribution Q_φ, we shall introduce a symbolic function $q(x)$ such that

$$Q_\varphi = \int q(x)\varphi(x) \ dx \ . \tag{6}$$

We shall say two symbolic functions $p(x)$ and $q(x)$ are equal if the corre-
sponding distributions $Q_\varphi = \int q(x) \, \varphi(x) \, dx$ and $P_\varphi = \int p(x) \, \varphi(x) \, dx$ are equal;
that is $p(x) = q(x)$ if and only if $P_\varphi = Q_\varphi$ for every testing function $\varphi(x)$. We
shall find it possible to interpret certain non-integrable functions such as
x^{-1} and $x^{-3/2}$ as distributions and thereby assign values to integrals which
would otherwise be regarded as divergent.

If α is any scalar, we define $\alpha q(x)$ as the symbolic function corresponding to the distribution αQ_φ. Also, $p(x) + q(x)$ is defined to be the
symbolic function corresponding to the distribution $P_\varphi + Q_\varphi$. In this way,
we can define linear combinations of symbolic functions.

DENSITY FUNCTIONS AND SYMBOLIC FUNCTIONS

We have seen that some distributions were defined by charge densities,
that is, that there existed an ordinary function $\rho(x)$ such that $Q_\varphi = \int \rho(x)\varphi(x)dx$.
In such cases we shall identify the symbolic function $q(x)$ corresponding to
$Q(x)$, with the ordinary function $\rho(x)$. This means that many symbolic functions will be identified with ordinary functions. In fact, suppose $f(x)$ is an
ordinary function of x, integrable over every finite interval, and of *slow
growth at infinity*, that is, there exists an integer n and a number C such
that $|f(x)| \le C|x|^n$ for all x of sufficiently large value; then the integral
$\int f(x)\varphi(x)dx$ exists for every testing function and defines a distribution.
We shall call functions such as $f(x)$ *density functions*. Since every density
function defines a distribution, we may identify every density function with
a symbolic function. But now, this identification of density functions with
symbolic functions raises the question of uniqueness. How do we know that
equality in the symbolic function sense is the same as equality in the ordinary sense? We must prove that if $f_1(x)$ and $f_2(x)$ are two density functions
such that $\int f_1(x) \, \varphi(x) \, dx = \int f_2(x) \, \varphi(x) \, dx$ for every testing function $\varphi(x)$
then $f_1(x) = f_2(x)$ in the ordinary sense. When we put $g(x) = f_1(x) - f_2(x)$,
this is equivalent to proving that if $g(x)$ is a density function such that
$\int g(x) \, \varphi(x) \, dx = 0$ for every testing function, the $g(x) = 0$. The proof of
this will be found in [Schwartz, 1950]*; consequently, for density functions,
equality in the symbolic sense is the same as equality in the ordinary sense.

MULTIPLICATION OF SYMBOLIC FUNCTIONS BY TESTING FUNCTIONS

In general, it is impossible to give a useful meaning to the product of
two symbolic functions. However, we can always assign a meaning to the
product of a symbolic function $s(x)$ by a testing function $\psi(x)$ by using the
equation

$$\int [s(x)\psi(x)] \, \varphi(x) \, dx = \int s(x)[\psi(x)\varphi(x)] \ dx \ .$$

The right-hand side has a meaning since the product of two testing functions
is again a testing function and it defines the left-hand side. For example,
this formula will be used to define $\psi(x)\delta(x)$. We have

---

* Names in brackets refer to publications cited in the bibliography at the end of the
  book.

$$\int [\delta(x)\psi(x)]\, \varphi(x)\, dx \;=\; \int \delta(x)[\psi(x)\varphi(x)]\, dx \;=\; \psi(0)\varphi(0) \;=\; \psi(0)\int \delta(x)\varphi(x)\, dx\,;$$

therefore

$$\psi(x)\delta(x) \;=\; \psi(0)\delta(x)\,.$$

This formula can be extended to any density function $f(x)$ which is continuous at the origin. We define $f(x)\delta(x) = f(0)\delta(x)$ or, more generally,

$$f(x)\delta(x - a) \;=\; f(a)\delta(x - a)\,. \tag{7}$$

However, other distributions than δ may require smoothness, etc., as well as continuity of f.

CHANGE OF VARIABLES

Our aim will be to develop a calculus of symbolic functions. The first concept we shall consider is that of change of variable in a symbolic function. Suppose that $s(x)$ is a symbolic function and suppose that $f(x)$ is a "good" function ("good" will be made explicit in context). What should be meant by $s[f(x)]$; for example, what should be meant by $\delta(3x - 4)$?

We shall give a meaning to the symbolic function $s[f(x)]$ by defining the value of the corresponding distribution, that is, the distribution whose values are given symbolically by $\int s[f(x)]\varphi(x)\, dx$. We define

$$\int s[f(x)]\, \varphi(x)\, dx \;=\; \int s(y)\left[\frac{d}{dy}\int_{f(\xi)<y}\varphi(\xi)\, d\xi\right] dy\,. \tag{8}$$

The right-hand side will have a meaning if $\dfrac{d}{dy}\int_{f(\xi)<y}\varphi(\xi)\, d\xi$ is a testing function. This will surely be the case if the inverse function to $f(x) = y$ is an analytic function of y of slow growth at infinity. For example, if $f(x) = \alpha x - \beta = y$, then $x = (y+\beta)/\alpha$, and (8) becomes

$$\int s(\alpha x - \beta)\varphi(x)\, dx \;=\; \int s(y)\left[\frac{d}{dy}\int_{\alpha\xi-\beta<y}\varphi(\xi)\, d\xi\right] dy\,. \tag{9}$$

Now put

$$\int_0^x \varphi(\xi)\, d\xi \;=\; \Phi(x);$$

then

$$\int_{\alpha\xi-\beta<y}\varphi(\xi)\, d\xi \;=\; \int_{-\infty}^{(y+\beta)/\alpha}\varphi(\xi)\, d\xi \;=\; \Phi\!\left(\frac{y+\beta}{\alpha}\right) - \Phi(-\infty) \quad \text{if } \alpha > 0,$$

$$=\; \int_{(y+\beta)/\alpha}^{\infty}\varphi(\xi)\, d\xi \;=\; \Phi(\infty) - \Phi\!\left(\frac{y+\beta}{\alpha}\right) \quad \text{if } \alpha < 0,$$

and

$$\frac{d}{dy}\int_{\alpha\xi-\beta<y}\varphi(\xi)\, d\xi \;=\; |\alpha|^{-1}\varphi[\alpha^{-1}(y+\beta)]$$

in all cases. Substituting this in (9), we find it becomes

$$\int s(\alpha x - \beta)\varphi(x)\,dx = |\alpha|^{-1} \int s(y)\varphi[\alpha^{-1}(y + \beta)]\,dy, \qquad (10)$$

the result that would have been obtained by a formal change of variable.

The particular case of (10) in which $s(y) = \delta(y)$ is important. In this case (10) becomes

$$\int \delta(\alpha x - \beta)\varphi(x)\,dx = |\alpha|^{-1} \int \delta(y)\varphi[\alpha^{-1}(y + \beta)]\,dy = |\alpha|^{-1}\varphi(\alpha^{-1}\beta). \qquad (11)$$

If $\beta = 0$, we find that

$$\delta(\alpha x) = |\alpha|^{-1}\delta(x); \qquad (12)$$

in particular, for $\alpha = -1$ we have $\delta(-x) = \delta(x)$, a result which permits us to say that $\delta(x)$ is an even function of x.

It is easy to show that if $f(x)$ is a monotonic function of x, that is, if $f'(x)$ has a constant sign, then formula (8) is equivalent to the usual formula for change of variables in an integral. However, formula (8) also makes sense in case $f(x)$ is not monotonic. For example, we shall use (8) to define $\delta(x^2 - a^2)$. We have

$$\int \delta(x^2 - a^2)\varphi(x)\,dx = \int \delta(y)\left[\frac{d}{dy} \int_{\xi^2-a^2<y} \varphi(\xi)\,d\xi\right]dy; \qquad (13)$$

but

$$\int_{\xi^2-a^2<y} \varphi(\xi)\,d\xi = \int_{-\sqrt{y+a^2}}^{+\sqrt{y+a^2}} \varphi(\xi)\,d\xi = \Phi(\sqrt{y + a^2}) - \Phi(-\sqrt{y + a^2}).$$

Therefore

$$\frac{d}{dy} \int_{\xi^2-a^2<y} \varphi(\xi)\,d\xi = \frac{1}{2\sqrt{y + a^2}}\left[\varphi(\sqrt{y + a^2}) + \varphi(-\sqrt{y + a^2})\right]$$

and the right-hand side of (13) becomes

$$\frac{1}{2}\int \delta(y)\frac{1}{\sqrt{y + a^2}}[\varphi(\sqrt{y + a^2}) + \varphi(-\sqrt{y + a^2})]\,dy = \frac{1}{2a}[\varphi(a) + \varphi(-a)].$$

Consequently, the symbolic function $\delta(x^2 - a^2)$ is defined by the formula

$$\int \delta(x^2 - a^2)\varphi(x)\,dx = \frac{1}{2a}[\varphi(a) + \varphi(-a)]. \qquad (14)$$

However, since the right-hand side of this can be written as

$$\frac{1}{2a}\int [\delta(x - a) + \delta(x + a)]\varphi(x)\,dx,$$

we conclude that

$$\delta(x^2 - a^2) = \frac{1}{2a}[\delta(x - a) + \delta(x + a)]. \qquad (15)$$

DERIVATIVES OF DISTRIBUTIONS

Since distributions are defined by integrals, it is natural to try to define derivatives of distributions by means of an integral formula containing derivatives. Such a formula is the one for integration by parts. If $f(x)$ is a continuous density function which has a piecewise continuous derivative and if $\varphi(x)$ is a testing function, then the formula for integration by parts gives

$$\int f'(x)\varphi(x)\ dx = f(x)\varphi(x)\Big|_{-\infty}^{\infty} - \int f(x)\varphi'(x)\ dx = -\int f(x)\varphi'(x)\ dx, \quad (16)$$

because $\varphi(x)$ decreases rapidly at infinity. We shall use (16) to define the derivative of a symbolic function.

Suppose $s(x)$ is a symbolic function, then, by definition, $s'(x)$ is the symbolic function for the distribution defined as follows:

$$\int s'(x)\varphi(x)\ dx = -\int s(x)\varphi'(x)\ dx. \quad (17)$$

As an illustration of this definition, consider $\delta'(x)$. From (17)

$$\int \delta'(x)\varphi(x)\ dx = -\int \delta(x)\varphi'(x)\ dx = -\varphi'(0);$$

therefore, $\delta'(x)$ picks out the negative of the value of the derivative of the testing function at the origin.

Since the right-hand side of (17) is defined for every symbolic function, it follows that every symbolic function has a symbolic function derivative. We shall call $s'(x)$ the symbolic derivative (sometimes just the derivative) of $s(x)$. Since $s'(x)$ is a symbolic function, it also has a derivative which we denote by $s''(x)$. Continuing this process, we see that every symbolic function will have derivatives of every order. For example, we find that

$$\int \delta^{(n)}(x)\varphi(x)\ dx = -\int \delta^{(n-1)}(x)\varphi'(x)\ dx = (-1)^n \int \delta(x)\varphi^{(n)}(x)\ dx$$
$$= (-1)^n \varphi^{(n)}(0); \quad (18)$$

this shows that $\delta^{(n)}(x)$ picks out the value of the nth derivative of $\varphi(x)$ at $x = 0$ multiplied by $(-1)^n$.

Every symbolic function has symbolic derivatives of all orders; consequently, every density function has symbolic derivatives of all orders. But what is the relationship between symbolic derivatives and ordinary derivatives? From (16), we see that if $f(x)$ is a continuous density function with a piecewise continuous derivative, then the ordinary derivative of $f(x)$ is the same as the symbolic derivative. Consider, however, the density function $H(x)$, the Heaviside unit function, which is defined by the formula

$$H(x) = \begin{cases} 1, & x > 0 \\ 0, & x < 0 \end{cases}.$$

For $x \neq 0$ the ordinary derivative of $H(x)$ is zero but for $x = 0$ the ordinary derivative of $H(x)$ does not exist. Using (16) to find $H'(x)$, the symbolic derivative of $H(x)$, we get $\int H'(x)\varphi(x)dx = -\int H(x)\varphi'(x)dx = -\int_0^\infty \varphi'(x)dx = \varphi(0)$ since $\varphi(x)$ vanishes for x infinite. Since $\varphi(0) = \int \delta(x)\varphi(x)dx$ we conclude that

$$H'(x) = \delta(x) . \tag{19}$$

Consider a piecewise continuous density function $f(x)$ which has a piece-wise continuous derivative. Suppose $f(x)$ has jumps of magnitude $a_1, a_2, \ldots, a_k$ at the values $x_1, x_2, \ldots, x_k$, respectively. Using the shifted Heaviside unit function $H(x - x_j)$, which is zero for $x < x_j$ and unity for $x > x_j$, we put

$$g(x) = f(x) - \sum_{j=1}^{k} a_j H(x - x_j) .$$

Note that $g(x)$ is continuous for all values of x and has a piecewise continuous derivative; therefore the symbolic derivative of $g(x)$ is the same as the ordinary derivative of $g(x)$. Since the ordinary derivative of the Heaviside function is zero wherever it exists, it follows that the ordinary derivative of $g(x)$ equals the ordinary derivative of $f(x)$ where both exist. Using the subscript s to denote symbolic derivative, we have

$$f'_s(x) = g'_s(x) + \sum_{j=1}^{k} a_j \delta(x - x_j) = g'(x) + \sum_{j=1}^{k} a_j \delta(x - x_j)$$

$$= f'(x) + \sum_{j=1}^{k} a_j \delta(x - x_j) . \tag{20}$$

This result shows that *the symbolic derivative of f(x) equals the ordinary derivative of f(x), wherever it exists, plus delta-functions at the places where f(x) has jumps multiplied by the magnitudes of the jumps*. Hereafter, we shall drop the subscript s and use the notation $f'(x)$ for the symbolic derivative.

Finally, we shall establish the chain rule for differentiating a symbolic function of a function. Suppose $s[f(x)]$, where $s(x)$ is a symbolic function and $f(x)$ is an ordinary function, is defined by (8). We shall prove that

$$\frac{d}{dx} s[f(x)] \equiv \{s[f(x)]\}' = s'[f(x)] f'(x) . \tag{21}$$

From the definition of the derivative in (17) and (8), we get

$$\int \{s[f(x)]\}' \varphi(x) \, dx = - \int s[f(x)] \varphi'(x) \, dx$$

$$= - \int s(y) \left[\frac{d}{dy} \int_{f(\xi) < y} \varphi'(\xi) d\xi \right] dy \tag{22}$$

$$= \int s'(y) \left[\int_{f(\xi) < y} \varphi'(\xi) d\xi \right] dy .$$

Suppose $f(\xi)$ is an increasing function of y and suppose that $f(\xi_0) = y$; then

$$\int_{f(\xi) < y} \varphi'(\xi) \, d\xi = \int_{-\infty}^{\xi_0} \varphi'(\xi) \, d\xi = \varphi(\xi_0) .$$

But note that if we put $\eta = f(\xi)$ then

$$\frac{d}{dy} \int_{f(\xi) < y} f'(\xi) \varphi(\xi) \, d\xi = \frac{d}{dy} \int_{-\infty}^{y} \varphi(\xi) \, d\eta = \varphi(\xi_0) ;$$

consequently,

$$\int_{f(\xi)<y} \varphi'(\xi)\,d\xi = \frac{d}{dy}\int_{f(\xi)<y} f'\,(\xi)\,\varphi\,(\xi)\,d\xi\,,$$

and (22) becomes

$$\int\{s[f(x)]\}'\,\varphi(x)\,dx = \int s'(y)\left[\frac{d}{dy}\int_{f(\xi)<y} f'(\xi)\varphi(\xi)d\ \,dy\right]$$

$$= \int s'\,[f(x)]\,f'(x)\,\varphi(x)\,dx$$

where the last equality follows from (8). This proves (21).

As an illustration of (21) suppose that $s(x) = H(x)$, the Heaviside unit function, and $f(x) = x^2 - a^2$; then from (21),

$$\frac{d}{dx}\,H\,(x^2 - a^2) = \delta(x^2 - a^2)\,2x = \frac{x}{a}[\delta(x - a) + \delta(x + a)]$$

$$= \delta(x - a) - \delta(x + a)\,,$$

(23)

by the use of (15) and (7). Note that the function $H(x^2 - a^2)$ is zero for $x^2 < a^2$ and is unity for $x < -a$ and $x > a$. This function has a jump of magnitude -1 at $x = -a$ and of magnitude $+1$ at $x = a$; consequently, its derivative is $-\delta(x + a) + \delta(x - a)$, which is the same as (23).

1.2 APPLICATIONS OF DISTRIBUTIONS

INTEGRATION OF DISTRIBUTIONS

It is well known that the integral of an ordinary function is unique except for the addition of an arbitrary constant. We shall show that the same fact is true for symbolic functions.

Suppose that $u(x)$ and $v(x)$ are symbolic functions which are both integrals of a symbolic function $s(x)$; then $u'(x) - v'(x) = s(x) - s(x) = 0$. Consequently, if we put $w(x) = u(x) - v(x)$, we have

$$w'(x) = 0\,.$$

(24)

We shall show that (24) implies $w(x)$ is a constant.

By the definition of the derivative of a symbolic function [see (17)], we have

$$0 = -\int w'(x)\,\varphi\,(x)\,dx = \int w(x)\,\varphi'(x)\,dx\,.$$

(25)

Note that the derivative $\varphi'(x)$ is itself a testing function and that $\int \varphi'(x)\,dx = \varphi(\infty) - \varphi(-\infty) = 0$. It is easy to verify that if, conversely, $\psi(x)$ is a testing function such that

$$\int \psi(x)\,dx = 0\,,$$

(26)

then the function $\varphi(x) = \int_{-\infty}^{x} \psi(\xi)\,d\xi$ is a testing function, and $\varphi'(x) = \psi(x)$.

Because of this, equation (25) shows that the symbolic function $w(x)$ is orthogonal to every testing function $\psi(x)$ that satisfies (26).

Let $\varphi_0(x)$ be a testing function such that $\int \varphi_0(x)\, dx = 1$, and let $\varphi_1(x)$ be an arbitrary testing function. Put $\int \varphi_1(x)\, dx = \alpha$, and put $\psi(x) = \varphi_1(x) - \alpha \varphi_0(x)$; then $\psi(x)$ is a testing function satisfying (26), and consequently

$$0 = \int w(x)\psi(x)\, dx = \int w(x)\, \varphi_1(x)\, dx - \alpha \int w(x)\, \varphi_0(x)\, dx .$$

This shows that, for any testing function $\varphi_1(x)$ we have $\int [w(x) - c]\, \varphi_1(x)\, dx = 0$, where $c = \int w(x)\, \varphi_0(x)\, dx$. Since the only symbolic function that is orthogonal to every testing function is the zero function, we conclude that the only solution of (24) is $w(x) = c$, a constant.

It is easy to generalize this result and to show that the only solution $w(x)$ of

$$w^{(n)}(x) = 0 \tag{27}$$

is a polynomial of degree less than n. Let us consider the nonhomogeneous equation

$$w^{(n)}(x) = \delta(x - a) . \tag{28}$$

We shall verify that a particular solution of this is

$$w_0(x) = \frac{(x - a)^{n-1}}{(n - 1)!} H(x - a) , \tag{29}$$

where $H(x - a)$ is the Heaviside unit function which is unity for $x > a$ and zero otherwise. We may also write (29) as

$$w_0(x) = \frac{(x - a)^{n-1}}{(n - 1)!} , \qquad x > a$$

$$= 0, \qquad\qquad x < a$$

From (20), we see that for $n > 1$,

$$w_0'(x) = \frac{(x - a)^{n-2}}{(n - 2)!} , \qquad x > a$$

$$= 0, \qquad\qquad x < a$$

because $w_0(x)$ is continuous at $x = a$ if $n > 1$. In the same way we find that, for $0 \le k < n$,

$$w_0^{(k)}(x) = \frac{(x - a)^{n-k-1}}{(n - k - 1)!} , \qquad x > a$$

$$= 0, \qquad\qquad x < a ;$$

but for $k = n$,

$$w_0^{(n)}(x) = \delta(x - a) .$$

Since the general solution of (28) is $w_0(x)$ plus a polynomial of degree less than n, we may describe the general solution as follows:

The general solution of (28) is an ordinary function of x which satisfies the homogeneous equation (27), for both $x < a$ and $x > a$. At $x = a$, the function and its first $(n - 2)$ derivatives are continuous, but the $(n - 1)$th derivative has a jump of magnitude unity. If $n = 1$, the function itself has a jump of magnitude one at $x = a$.

Similar results hold for the solution of the nth order differential equation

$$\frac{d^n u}{dx^n} + p_1(x) \frac{d^{n-1} u}{dx^{n-1}} + \cdots + p_n(x)u = \delta(x - a) \tag{30}$$

if the functions $p_1(x), \ldots, p_n(x)$ are continuous at $x = a$. We can show that, for both $x < a$ and $x > a$, the symbolic function $u(x)$ is the solution of the homogeneous equation

$$\frac{d^n u}{dx^n} + p_1(x) \frac{d^{n-1} u}{dx^{n-1}} + \cdots + p_n(x)u = 0 \tag{31}$$

and that for $x = a$, the function $u(x)$ and its first $(n-2)$ derivatives are continuous but its $(n-1)$th derivative has a jump of magnitude unity there.

To illustrate the methods used to solve an equation such as (30), we consider the problem of finding the steady-state displacement of a string fixed at $x = 0$ and $x = 1$ if it is driven by a simple-harmonic force of radian frequency ω concentrated at a point $x = \xi$. The equation of motion of the vibrating string is

$$u_{xx} - \frac{1}{c^2} u_{tt} = F \tag{32}$$

where c is the velocity of sound on the string and F is proportional to the impressed force. In this case, we take $F = \delta(x - \xi)e^{-i\omega t}$. Put $u = v(x)e^{-i\omega t}$ so that (32) becomes

$$v'' + K^2 v = \delta(x - \xi) \tag{33}$$

where $K^2 = \dfrac{\omega^2}{c^2}$. Since the string is fixed at $x = 0$ and $x = 1$, we must have

$$v(0) = v(1) = 0. \tag{34}$$

To solve (33) with the boundary conditions (34), we proceed as follows: For $x < \xi$ the solution of the homogeneous equation corresponding to (33), namely,

$$v'' + K^2 v = 0, \tag{35}$$

that satisfies the boundary condition $v(0) = 0$ is $v = \sin Kx$. For $x > \xi$, the solution of (35) that satisfies the boundary condition $v(1) = 0$ is $v = \sin K(1 - x)$. Therefore, we have

$$v = A \sin Kx, \qquad x < \xi$$

$$= B \sin K (1 - x), \qquad x > \xi$$

where A and B are constants to be determined by the conditions at $x = \xi$. We require that $v(x)$ be continuous for $x = \xi$ and that its first derivative have a jump of magnitude one at $x = \xi$. These conditions imply that

$$A \sin K \xi - B \sin K(1-\xi) = 0, \qquad - KA \cos K \xi - KB \cos K(1 - \xi) = 1.$$

When these equations are solved, we find that

$$A = - \frac{\sin K(1 - \xi)}{KD},$$

$$B = - \frac{\sin K\xi}{KD},$$

$$D = \sin K\xi \cos K(1 - \xi) + \sin K(1 - \xi) \cos K\xi = \sin K,$$

and consequently

$$v = - \frac{\sin Kx \sin K(1 - \xi)}{K \sin K}, \qquad x < \xi$$

$$= - \frac{\sin K(1 - x) \sin K\xi}{K \sin K}, \qquad x > \xi.$$

Notice that $v(x,\xi) = v(\xi,x)$, that is, that the function $v(x,\xi)$ is unchanged when x and ξ are interchanged. This is an example of a reciprocity theorem: The displacement produced at x by a force located at ξ is equal to the displacement produced at ξ by a force located at x.

We conclude by finding the displacement produced on a string of infinite length by a simple harmonic force concentrated at $x = \xi$. The equation for the displacement is again

$$v'' + K^2 v = \delta(x - \xi), \tag{36}$$

but in this case the required boundary conditions at $\pm\infty$ are that $v(x)$ be an "outgoing" wave, that is, that $v(x)e^{-i\omega t}$ be a wave progressing outward from the force at $x = \xi$. With these conditions it is easy to verify that the solution of (36) is $v(x,\xi) = e^{iK|x-\xi|}/2iK$. [See *Friedman*, Ch. 3 for additional discussion of the "Green's function" $v(x,\xi)$.]

LIMITS OF SYMBOLIC FUNCTIONS

Suppose that $s_n(x)$, $(n = 1, 2, 3, \ldots)$ is a sequence of symbolic functions. The sequence is said to converge to a symbolic function $s(x)$ as a limit if, for every testing function $\varphi(x)$,

$$\lim_{n \to \infty} \int s_n(x) \varphi(x) \, dx = \int s(x) \varphi(x) \, dx. \tag{37}$$

By an obvious extension of this definition, the symbolic functions $s(x,\epsilon)$, depending on a parameter ϵ, are said to converge to the symbolic function $s(x)$ as ϵ approaches zero if, for every testing function $\varphi(x)$,

$$\lim_{\epsilon \to 0} \int s(x,\epsilon)\varphi(x)\, dx = \int s(x)\varphi(x)\, dx. \qquad (38)$$

As an illustration of this definition, we show that the limit as ϵ approaches zero of the symbolic functions $\epsilon/(x^2 + \epsilon^2)$ is $\pi\delta(x)$. We have

$$I(\epsilon) = \int \frac{\epsilon}{x^2 + \epsilon^2}\, \varphi(x)\, dx$$

$$= \varphi(0) \int \frac{\epsilon\, dx}{x^2 + \epsilon^2} + \int \left[\varphi(x) - \varphi(0)\right] \frac{\epsilon}{x^2 + \epsilon^2}\, dx.$$

Put $x = \epsilon y$; then

$$I(\epsilon) = \varphi(0) \int \frac{dy}{1 + y^2} + \int \left[\varphi(\epsilon y) - \varphi(0)\right] \frac{dy}{1 + y^2}.$$

Since

$$\int \frac{dy}{1 + y^2} = \pi,$$

we shall have $\lim_{\epsilon \to 0} I(\epsilon) = \pi\varphi(0) = \pi \int \delta(x)\varphi(x)\, dx$ if we can prove that the term containing $[\varphi(\epsilon y) - \varphi(0)]$ converges to zero as ϵ approaches zero. The proof of this is as follows.

Since $\varphi(x)$ is a testing function it is bounded; therefore there exists a constant M such that $|\varphi(\epsilon y) - \varphi(0)| < M$ for all values of y. If n is any positive integer, pick $\eta > 4nM$; then

$$\left| \int_{|y|>\eta} \left[\varphi(\epsilon y) - \varphi(0)\right] \frac{dy}{1+y^2} \right| < M \int_{|y|>\eta} \frac{dy}{1 + y^2} < 2M \int_{\eta}^{\infty} \frac{dy}{y^2} < \frac{1}{2n}.$$

Since $\varphi(x)$ is continuous, we can find ϵ_0 such that for $0 < \epsilon < \epsilon_0$ we have $|\varphi(\epsilon y) - \varphi(0)| < (2\pi n)^{-1}$ for all values of y such that $|y| \leq \eta$; then

$$\left| \int_{|y|\leq\eta} \left[\varphi(\epsilon y) - \varphi(0)\right] \frac{dy}{1+y^2} \right| < (2\pi n)^{-1} \int \frac{dy}{1 + y^2} = \frac{1}{2n}.$$

Thus we have shown

$$\left| \int \left[\varphi(\epsilon y) - \varphi(0)\right] \frac{dy}{1 + y^2} \right| \leq \left| \int_{|y|\leq\eta} \right| + \left| \int_{|y|>\eta} \right| \leq \frac{1}{n};$$

this proves that the limit of $I(\epsilon)$ is $\pi\varphi(0)$ and that

$$\lim_{\epsilon \to 0} \epsilon (x^2 + \epsilon^2)^{-1} = \pi\delta(x). \qquad (39)$$

With the definition of limit given in (38), it is easy to show that, for any symbolic function $s(x)$,

$$\lim_{h \to 0} \frac{s(x+h) - s(x)}{h} = s'(x), \qquad (40)$$

where $s'(x)$ is defined by (17). Because of (38), it is sufficient to show that for every testing function

$$\lim_{h \to 0} \int \left[\frac{s(x+h) - s(x)}{h} \right] \varphi(x)\, dx = \int s'(x)\varphi(x)\, dx.$$

By a change of variables, we have

$$\frac{1}{h} \int [s(x + h) - s(x)]\, \varphi(x)\, dx = \frac{1}{h} \int s(x)[\varphi(x - h) - \varphi(x)]\, dx,$$

and since

$$\lim_{h \to 0} \frac{\varphi(x - h) - \varphi(x)}{h} = -\varphi'(x),$$

we obtain

$$\lim_{h \to 0} \int s(x) \left[\frac{\varphi(x - h) - \varphi(x)}{h} \right] dx = - \int s(x)\varphi'(x)\, dx.$$

But by (17), $\int s(x)\,\varphi'(x)\, dx = -\int s'(x)\,\varphi(x)\, dx$; therefore (40) is established.

We now prove the following theorem:

Theorem: If the symbolic functions $s(x,\epsilon)$ converge to the limit $s(x)$ as ϵ approaches zero then the derivatives $s'(x,\epsilon)$ converge to the derivative $s'(x)$.

From (17), for every testing function we have

$$\int s'(x,\epsilon)\varphi(x)\, dx = - \int s(x,\epsilon)\varphi'(x)\, dx.$$

By assumption, the right-hand side converges to

$$- \int s(x)\varphi'(x)\, dx = \int s'(x)\varphi(x)\, dx$$

and therefore $s'(x,\epsilon)$ converges to $s'(x)$.

SYMBOLIC FUNCTIONS IN MORE THAN ONE DIMENSION

The introduction of symbolic functions was motivated by the need to obtain a volume charge density for charge distributions defined by point charges, dipoles, or other charge distributions which do *not* have a volume charge density. We shall see that in such cases the volume charge density may be represented by a symbolic function.

We begin by defining testing functions in three dimensions. The function $\varphi(\vec{r})$ is a testing function if it is continuous and has continuous derivatives of all orders for all values of $\vec{r}$ and if, for every pair of integers n and m, there exists a constant M such that

$$\left| r^n\, \frac{\partial^{m_1}}{\partial x^{m_1}} \frac{\partial^{m_2}}{\partial y^{m_2}} \frac{\partial^{m_3}}{\partial z^{m_3}}\, \varphi(\vec{r}) \right| < M$$

where $m = m_1 + m_2 + m_3$. If Q_φ is a linear continuous functional on the space of testing functions, then Q_φ is called a distribution; we introduce a symbolic function $q(\vec{r})$, and write $Q_\varphi = \int q(\vec{r})\varphi(\vec{r})\,d\vec{r}$ where the integration is extended over all of three-dimensional space. For example, a unit point charge at the point $\vec{r} = \vec{r}_0$ would be represented by the symbolic function $\delta(\vec{r} - \vec{r}_0)$, since $\int \delta(\vec{r} - \vec{r}_0)\varphi(\vec{r})dr = \varphi(\vec{r})$. But, if we introduce rectangular coordinates,

$$\iiint \delta(x - x_0)\delta(y - y_0)\delta(z - z_0)\varphi(x, y, z)\, dx\, dy\, dz = \varphi(x_0 y_0 z_0);$$

therefore

$$\delta(\vec{r} - \vec{r}_0) = \delta(x - x_0)\delta(y - y_0)\delta(z - z_0). \tag{41}$$

It is now easy to see that a symbolic function such as $\delta(x - x_0)\delta(y - y_0)$ represents a uniform *line charge* along the straight line $x = x_0$, $y = y_0$, and a symbolic function such as $\delta(x - x_0)$ represents a uniform *surface charge* along the plane $x = x_0$.

Similar representations can be given for the current density vector $\vec{J}$ in Maxwell's equations. For example, $\delta(x - x_0)\vec{1}_y$ where $\vec{1}_y$ is a unit vector in the y-direction, represents a uniform *surface current* of charge on the plane $x = x_0$ moving in the y direction. The symbolic function

$$\delta(x - x_0)\,\delta(y - y_0)\vec{1}_z$$

represents a *line current* of charge on the line $x = x_0$, $y = y_0$ moving in the z direction. Finally, the symbolic function

$$\delta(x - x_0)\delta(y - y_0)\delta(z - z_0)\vec{1}_z$$

represents a *point current*, a point charge moving in the z direction.

By an obvious extension of (17), we define the derivative $(\vec{a}\cdot\nabla)g(\vec{r})$ as follows:

$$\int [(\vec{a}\cdot\nabla)g(\vec{r})]\,\varphi(\vec{r})\, d\vec{r} = - \int g(\vec{r})(\vec{a}\cdot\nabla)\varphi(\vec{r})\, d\vec{r}, \tag{42}$$

where $\vec{a}$ is any constant vector. In particular, (42) implies that

$$\iiint \frac{\partial}{\partial x} g(x, y, z)\varphi(x, y, z)\, dx\, dy\, dz = - \iiint g\frac{\partial\varphi}{\partial x}\, dx\, dy\, dz.$$

Just as in the preceding section on limits of symbolic functions, it is easy to show that

$$\lim_{h\to 0} \frac{g(x + h, y, z) - g(x, y, z)}{h} = \frac{\partial g(x, y, z)}{\partial x}, \tag{43}$$

and that if the symbolic functions $g(\vec{r}, \epsilon)$ converge to a symbolic function $g(\vec{r})$ as ϵ approaches zero, then any derivative of $g(\vec{r}, \epsilon)$ converges to the corresponding derivative of $g(\vec{r})$.

With the help of (43) we shall obtain a representation for a dipole charge. A dipole of moment μ, at the point $x = x_0$, $y = y_0$, $z = z_0$ with axis along the positive x direction, is the limit as h approaches zero of a negative charge of magnitude μ/h, at the point $x = x_0$, $y = y_0$, $z = z_0$, combined with a positive charge of magnitude μ/h, at the point $x = x_0 + h$, $y = y_0$, $z = z_0$. The volume charge density produced by such a combination of charges is

$$\frac{\mu}{h}\left[-\delta(x - x_0)\delta(y - y_0)\delta(z - z_0) + \delta(x - x_0 - h)\delta(y - y_0)\delta(z - z_0)\right]$$

and in the limit, as h approaches zero, it becomes

$$\mu\delta(y - y_0)\delta(z - z_0)\lim\frac{\delta(x - x_0 - h) - \delta(x - x_0)}{h}$$

$$= -\mu\delta'(x - x_0)\delta(y - y_0)\delta(z - z_0).$$

Similarly, a quadrupole defined as the limit of positive and negative dipoles with axes along the x axis displaced an infinitesimal distance along the x axis would have a volume charge density such as

$$\delta''(x - x_0)\delta(y - y_0)\delta(z - z_0).$$

If the dipoles were displaced an infinitesimal distance along the y axis, the volume charge density would be

$$\delta'(x - x_0)\delta'(y - y_0)\delta(z - z_0).$$

The extensions to multipoles of higher order are clear.

A volume charge density such as $\delta'(x - x_0)\delta(y - y_0)$ represents a straight line $(x = x_0, y = y_0)$ of *dipoles* with axes along the x axis, whereas $\delta'(x - x_0)$ represents a plane $(x = x_0)$ *layer of dipoles* with axes along the x axis.

ILLUSTRATION FROM ELECTROSTATICS

Given a volume charge density ρ, the electric field intensity $\vec{E}$ satisfies the equation

$$\nabla\cdot\vec{E} = \frac{\rho}{\epsilon} \tag{44}$$

where ϵ is the electric permittivity of the medium. In a static problem, Maxwell's equations state that $\nabla\times\vec{E} = 0$, and therefore $\vec{E}$ can be obtained from a potential function ψ; that is,

$$\vec{E} = -\nabla\psi. \tag{45}$$

Combining (44) and (45), we get

$$\nabla^2\psi = \frac{\partial^2\psi}{\partial x^2} + \frac{\partial^2\psi}{\partial y^2} + \frac{\partial^2\psi}{\partial z^2} = -\frac{\rho}{\epsilon}. \tag{46}$$

The relations (44)—(46) have been derived on the assumption that the volume charge density ρ was an ordinary function. However, it can be shown that these relations will still be valid when ρ is a symbolic function if the derivatives are symbolic derivatives. Using this fact, we can obtain information about the continuity properties of the potential and the electric intensity.

Suppose, first that ρ is an ordinary function; then ψ and all its derivatives must be continuous. For, otherwise, the second derivatives of ψ in (46) would contain δ function or δ' terms. This implies that a volume charge density produces a continuous potential and a continuous electric field intensity.

Second, suppose there is a surface charge density on the plane $z = 0$. This would be represented by $\rho = \rho_S(x, y)\delta(z)$. In order to get such a δ function in (46), the function ψ must be continuous, but $\partial\psi/\partial z$ must have a jump of magnitude $-\rho_S/\epsilon$ at $z = 0$. This implies that a surface charge produces a continuous potential and, from (45), a jump in the normal component of the electric field intensity across the surface charge.

Finally, suppose there is a layer of dipoles on the plane $z = 0$. This would be represented by $\rho = \rho_d(x, y)\delta'(z)$. In order to get a $\delta'(z)$ in (46), the function ψ must have a jump of magnitude ρ_d/ϵ at $z = 0$, and $\partial\psi/\partial z$ $+(\rho_d/\epsilon)\delta(z)$ be continuous across $z = 0$. This implies that a surface layer of dipoles produces at the layer a discontinuous potential, discontinuous tangential components of electric field intensity, and an "infinite" normal component of electric field intensity (a distribution).

EQUATION OF CONTINUITY FOR ELECTRIC CHARGE

According to Maxwell's theory, if $\vec{J}$ is the electric current density vector, and if ρ is the electric charge density, then

$$\nabla \cdot \vec{J} + \partial\rho/\partial t = 0. \tag{47}$$

Again, this relation will be true for symbolic functions and symbolic derivatives. We shall give two illustrations of the use of (47).

First, consider the case in which both ρ and $\vec{J}$ vary simply harmonically with time as $e^{-i\omega t}$, and suppose that ρ is just a point dipole $\delta(x)\delta(y)\delta'(z)$. Then equation (47) becomes $\nabla \cdot \vec{J} = i\omega\delta(x)\delta(y)\delta'(z)e^{-i\omega t}$. A solution of this equation is

$$\vec{J} = i\omega\delta(x)\delta(y)\delta(z)\vec{1}_z\, e^{-i\omega t} \tag{48}$$

where $\vec{1}_z$ is a unit vector in the z direction. We see that a point current density such as (48) can be interpreted as the current produced by an oscillating point dipole charge.

The second case we shall consider is that of a finitely conducting sphere of radius a in which a volume charge density $\rho_0(r)$ is placed at time $t = 0$. What happens to the charge? If we let σ be the conductivity of the sphere, then $\vec{J} = \sigma\vec{E}$. Using this equation, together with (47) and (44), we get

$$\partial\rho/\partial t = -\nabla \cdot (\sigma\vec{E}) = -\sigma\nabla \cdot \vec{E} = -\sigma\rho/\epsilon$$

The solution of this equation is $\rho = \rho_0(r)e^{-\sigma t/\epsilon}$, which implies that the total charge inside the sphere decreases to zero as time increases. However, by the conservation of charge, we know that the decrease in charge must appear somewhere, and on physical grounds we expect it to appear as surface charge on the sphere. We shall see that the mathematics gives this result if symbolic functions are used.

We assume that all the quantities ρ, $\vec{E}$, $\vec{J}$ depend only on r and t, and we assume that both the electric permittivity of the sphere and the medium surrounding it have the same value ϵ. From (47) we get $\nabla \cdot (\sigma \vec{E}) + \partial\rho/\partial t = 0$. Since $\sigma = 0$ for $r > a$, we see that σE is discontinuous at $r = a$ and therefore its derivative must contain a δ function. This immediately implies that ρ must contain a δ-function term, that is, a surface charge. Since the only dependence is on r and t, it is sufficent to assume $\vec{E}$ has only an r component. Let E_1 denote this r component for $r < a$, and let E_2 denote this r component for $r > a$. Put ρ_1 equal to the volume charge density inside the sphere, ρ_2 equal to the volume charge density outside, and $\rho_s \delta(r - a)$ equal to the surface charge density. Since $\nabla \cdot A(r)\vec{1}_r = (1/r^2)(\partial/\partial r)(r^2 A_r)$, and since at $r = a$ we have $\nabla \cdot [\vec{E}\sigma H(a - r)] = -\sigma E_1 \delta(r-a) = -(\partial\rho_s/\partial t)\delta(r-a)$, we find

$$\frac{\sigma}{r^2}\frac{\partial(r^2 E_1)}{\partial r} + \frac{\partial\rho_1}{\partial t} = 0; \qquad \frac{\sigma}{r^2}\frac{\partial(r^2 E_2)}{\partial r} + \frac{\partial\rho_2}{\partial t} = 0, \ \sigma = 0;$$

$$-\sigma E_1|_a + (\partial\rho_s/\partial t) = 0, \tag{49}$$

where $\sigma E_1|_a$ is the value of σE_1 at $r = a$. From (44),

$$\frac{1}{r^2}\frac{\partial(r^2 E_1)}{\partial r} = \frac{\rho_1}{\epsilon}, \quad \frac{1}{r^2}\frac{\partial(r^2 E_2)}{\partial r} = \frac{\rho_2}{\epsilon}. \tag{50}$$

Substituting (50) into (49), we get $\partial\rho_1/\partial t = -(\sigma/\epsilon)\rho_1$, and $\partial\rho_2/\partial t = 0$. The solution of these equations is $\rho_1(r, t) = \rho_0(r)e^{-\sigma t/\epsilon}$, and $\rho_2(r, t) = 0$ (since at $t = 0$, $\rho_2 = 0$). From the first equation of (50) we get

$$r^2 E_1|_a = \frac{1}{\epsilon}\int_0^a r^2 \rho_0(r)dr\, e^{-\sigma t/\epsilon} \equiv (Qe^{-\sigma t/\epsilon})/4\pi\epsilon.$$

Using this in the last equation of (49) and remembering that $\rho_s = 0$ at $t = 0$, we find that $\rho_s = (Q/4\pi a^2)(1 - e^{-\sigma t/\epsilon})$. Since the total volume charge in the sphere is $4\pi\int_0^a \rho_1(r)r^2\, dr = Qe^{-\sigma t/\epsilon}$ and the total surface charge is $4\pi a^2 \rho_s$, we see that at all times the sum of the volume charge plus the surface charge is $Q = \int_0^a 4\pi r^2 \rho_0(r)\, dr$, the initial total charge on the sphere.

1.3 DISTRIBUTIONS AND TRANSFORMS

AN INTEGRAL REPRESENTATION FOR $\delta(x)$

Before discussing the Fourier integral and the Laplace transform, we shall derive the following important integral representation for the delta function:

$$\delta(x) = \frac{1}{2\pi}\int_{-\infty}^{\infty} e^{ikx}\, dk. \tag{51}$$

We begin the proof of this result by considering the well-known integral

$$I = \int_{-\infty}^{\infty} \frac{e^{ikx}}{k}\, dk \tag{52}$$

which involves two implicit procedures. Because of the infinite behavior of the integrand in (52) for $k = 0$, the integral must be considered as a principal-value integral, that is,

$$I = \lim_{\epsilon \to 0} \left[\int_{-\infty}^{-\epsilon} + \int_{\epsilon}^{\infty} \right] \frac{e^{ikx}}{k}\, dk .$$

Also, the infinite limits are to be interpreted as the limit of the integral between $(-R, R)$ as R goes to infinity, so that, to be completely explicit, we should write

$$I = \lim_{\substack{\epsilon \to 0 \\ R \to \infty}} \left[\int_{-R}^{-\epsilon} + \int_{\epsilon}^{R} \right] \frac{e^{ikx}}{k}\, dk . \tag{53}$$

The integral in (53) can be easily evaluated by contour integration. If $x > 0$, we consider the contour in the k-plane as in Figure 1.1:

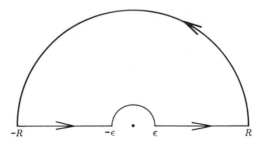

Figure 1.1

If $x < 0$, we consider a similar contour but with the semicircles in the lower half-plane. Since the integrand e^{ikx}/k has no singularity inside the contour in the above diagram, we have by Cauchy's theorem

$$\left[\int_{-R}^{-\epsilon} + \int_{\epsilon}^{R} + \int_{C_\epsilon} + \int_{C_R} \right] \frac{e^{ikx}}{k}\, dk = 0 , \tag{54}$$

where C_ϵ and C_R denote the semicircles of radius ϵ and R, respectively, To evaluate the integrals around C_ϵ and C_R, put $k = \epsilon e^{i\theta}$ and $k = R e^{i\theta}$, respectively. We get

$$\int_{C_\epsilon} = i \int_\pi^0 \exp\Big(x[i\epsilon\cos\theta - \epsilon\sin\theta]\Big)\,d\theta,$$

and then

$$\lim_{\epsilon\to 0} \int_{C_\epsilon} = i \int_\pi^0 d\theta = -i\pi. \qquad (55)$$

We have, similarly, that

$$\int_{C_R} = i \int_0^\pi \exp x[iR\cos\theta - R\sin\theta]\,d\theta.$$

Since $|\exp(iR\cos\theta)| \le 1$ we see that

$$\left|\int_{C_R}\right| \le \int_0^\pi e^{-xR\sin\theta}\,d\theta = 2\int_0^{\frac{\pi}{2}} e^{-xR\sin\theta}\,d\theta.$$

However, if we compare the graph of $\sin\theta$ for $0 \le \theta \le \pi/2$ with the graph of $2\theta/\pi$ in Figure 1.2,

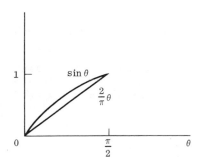

Figure 1.2

we see that $\sin\theta \ge 2\theta/\pi$, $0 \le \theta \le \pi/2$; consequently,

$$0 \le \int_0^{\frac{\pi}{2}} e^{-xR\sin\theta}\,d\theta \le \int_0^{\frac{\pi}{2}} \exp\left[-\frac{2R}{\pi}\theta\right]\,d\theta = \frac{\pi}{2R}\,(1-e^{-R})\to 0,$$

as $R \to \infty$. This shows that

$$\lim_{R\to\infty} \int_{C_R} = 0. \qquad (56)$$

Combining (54)–(56), we get

$$I = \lim_{\substack{\epsilon \to 0 \\ R \to \infty}} \left[\int_{-R}^{-\epsilon} + \int_{\epsilon}^{R} \right] \frac{e^{ikx}}{k} \, dk = i\pi. \tag{57}$$

When a similar procedure is applied in the case $x < 0$, we find that $I = -i\pi$. Both results can be combined as

$$I = \text{P.V.} \int_{-\infty}^{\infty} \frac{e^{ikx}}{k} \, dk = \pi \, \text{sgn} \, x, \tag{58}$$

where P.V. (which we usually drop) stands for principal value, and

$$\text{sgn} \, x = +1, \quad x > 0$$
$$= -1, \quad x < 0. \tag{59}$$

If we separate (56) into real and imaginary parts, we get the following useful integrals:

$$\int_{-\infty}^{\infty} \frac{\cos kx}{k} \, dk = 0,$$

$$\int_{-\infty}^{\infty} \frac{\sin kx}{k} \, dk = 2 \int_{0}^{\infty} \frac{\sin kx}{k} \, dk = \pi \, \text{sgn} \, x. \tag{60}$$

Since the function sgn x is continuously differentiable, except for a jump in magnitude of 2 at $x = 0$, we have $(\text{sgn} \, x)' = 2\delta(x)$. Differentiating with respect to x and using this result, we obtain formula (51) in the form $\int_{-\infty}^{\infty} e^{ikx} \, dk = 2\pi\delta(x)$.

THE FOURIER INTEGRAL

From formula (51) we can derive the Fourier integral theorem. Replace x by $x - y$ in (51), thus getting the formula $\delta(x - y) = (1/2\pi) \int e^{ik(x-y)} dk$. Multiply both sides of this equation by a testing function $\varphi(y)$ and integrate. We find that $\int \delta(x - y)\varphi(y)dy = \varphi(x)$ and $(1/2\pi) \int \varphi(y)dy \int e^{ik(x-y)} \, dk = (1/2\pi) \int e^{-ikx} \, dk \int e^{-iky} \varphi(y)dy$; therefore

$$\varphi(x) = \frac{1}{2\pi} \int e^{ikx} \, dk \int e^{-iky} \varphi(y) \, dy, \tag{61}$$

which is the Fourier integral theorem. Note all integrals are from $-\infty$ to $+\infty$.

Put

$$\psi(k) = \int e^{-iky} \varphi(y) \, dy, \tag{62}$$

then

$$\varphi(x) = \frac{1}{2\pi} \int \psi(k) e^{ikx} \, dk \, . \tag{63}$$

We call $\psi(k)$ the *Fourier transform* of $\varphi(y)$. It can be shown that every distribution has a Fourier transform. For example, from (51) the Fourier transform of the function that is equal to one for all values of x is $2\pi\delta(x)$. We may write this as $\int e^{-ikx} dx = 2\pi\delta(k)$ [remember $\delta(-k) = \delta(k)$]. Also, since $\int \delta(x) e^{-ikx} \, dx = 1$, the Fourier transform of $\delta(x)$ is the constant function equal to one.

To find the Fourier transform of $H(x)$, we begin with the integral $I_1 = \int_0^\infty e^{-\epsilon x} e^{-ikx} \, dx = (\epsilon + ik)^{-1}$, $\epsilon > 0$. We use

$$\lim_{\epsilon \to 0} \frac{1}{\epsilon + ik} = \lim_{\epsilon \to 0} \frac{\epsilon - ik}{\epsilon^2 + k^2} = \pi\delta(k) - i \text{ P.V.} \frac{1}{k},$$

where the first term ($\pi\delta$) follows from (39), and where the second [essentially as in (58)] will be discussed subsequently. Thus

$$\lim_{\epsilon \to 0} I_1 = \int H(x) e^{-ikx} \, dx = \pi\delta(k) - i \text{ P.V.} \frac{1}{k} \tag{64}$$

If we separate real and imaginary parts in (64), we get

$$\int_0^\infty \cos kx \, dx = \pi\delta(k), \tag{65}$$

$$\int_0^\infty \sin kx \, dx = \text{P.V.} \frac{1}{k} \tag{66}$$

In the following, in general, we use $1/k$ as a symbolic function for the distribution whose value is given by P.V. $\int(\varphi(k)/k) \, dk$.

Let us compare these results with those obtained by taking real and imaginary parts of (51). We find that

$$\int_{-\infty}^\infty \cos kx \, dx = 2\pi\delta(k), \tag{67}$$

$$\int_{-\infty}^\infty \sin kx \, dx = 0. \tag{68}$$

Note that since the cosine is an even function of x, the right-hand side of (67) is twice the right-hand side of (65). Also, since the sine is an odd function of x, the result in (68) is zero. The result of (66) can be remembered easily if we notice that $\int \sin kx \, dx = -(\cos kx)/k$, and this evaluated at the lower limit $x = 0$ gives the correct symbol.

EXAMPLES OF FOURIER TRANSFORMS

Further examples of the Fourier transform may be obtained by differentiating formulas (65) and (66) n times with respect to k. In this way we obtain the results

$$\int_0^\infty x^{2n} \cos kx \; dx = \pi(-1)^n \delta^{(2n)}(k),$$

$$\int_0^\infty x^{2n} \sin kx \; dx = (-1)^n (2n)! \; k^{-(2n+1)},$$

$$\int_0^\infty x^{2n+1} \sin kx \; dx = \pi(-1)^{n+1} \delta^{(2n+1)}(k),$$

$$\int_0^\infty x^{2n+1} \cos kx \; dx = (-1)^{n+1} (2n+1)! k^{-(2n+2)}.$$

(69)

These integrals can easily be extended to $(-\infty, \infty)$ by using the properties that $\cos kx$ is an even and $\sin kx$ is an odd function of x. We find

$$\int_{-\infty}^\infty x^{2n} \cos kx \; dx = 2\pi (-1)^n \delta^{(2n)}(k),$$

$$\int_{-\infty}^\infty x^{2n+1} \sin kx \; dx = 2\pi (-1)^{n+1} \delta^{(2n+1)}(k),$$

$$\int_{-\infty}^\infty x^{2n} \operatorname{sgn} x \; \sin kx \; dx = 2(-1)^n (2n)! k^{-(2n+1)},$$

$$\int_{-\infty}^\infty x^{2n+1} \operatorname{sgn} x \; \cos kx \; dx = 2(-1)^{n+1} (2n+1)! \; k^{-(2n+2)}.$$

(70)

These results can easily be transformed into corresponding integrals containing e^{ikx}. A suitable combination of these formulas gives the following:

$$\int_{-\infty}^\infty x^n e^{ikx} \; dx = 2\pi (-i)^n \delta^{(n)}(k),$$

(71)

$$\int_{-\infty}^\infty x^n \operatorname{sgn} x \; e^{ikx} \; dx = 2i^{n+1} n! \; k^{-(n+1)}.$$

(72)

Note that formulas (71) and (72) contain all of (69) and (70).

It is easy to verify that the inverse Fourier transform of the right-hand side of (71) is x^n. We have $(1/2\pi) \int e^{-ikx} [2\pi (-i)^n \delta^{(n)}(k)] dk = x^n$, as required.

A similar verification for (72) will follow once we define the meaning of integrals such as $\int(e^{-ikx}/k^{n+1})dk$.

The formulas (71) and (72) are valid only for $n = 0, 1, 2, \ldots$. We shall obtain similar formulas for n having the values $-\frac{1}{2}, \frac{1}{2}, \frac{3}{2}, \ldots$. We start with the well-known Fresnel integrals:

$$\int_0^\infty \frac{\cos kx\, dx}{x^{1/2}} = \int_0^\infty \frac{\sin kx\, dx}{x^{1/2}} = \sqrt{\frac{\pi}{2k}} \tag{73}$$

[See Erdélyi, et al. for all "well known" results and special functions we touch on.] These can be combined to give

$$\int_0^\infty \frac{e^{ikx}}{x^{1/2}}\, dx = \left(\frac{\pi}{k}\right)^{1/2} e^{\pi i/4}, \tag{74}$$

where, for $k < 0$, we must put $k^{1/2} = i|k|^{1/2}$. We prefer to write (74) as follows:

$$\int_{-\infty}^\infty x^{-1/2} H(x)\, e^{ikx}\, dx = \left(\frac{\pi}{k}\right)^{1/2} e^{\pi i/4}. \tag{75}$$

Differentiating this formula n times with respect to k, we get

$$\int_{-\infty}^\infty x^{n-1/2} H(x)\, e^{ikx}\, dx = \Gamma\left(n+\frac{1}{2}\right) e^{(n+1/2)\pi i/2} k^{-(n+1/2)}, \tag{76}$$

where Γ is the gamma function such that $\Gamma(z+1) = z\,\Gamma(z)$, $\Gamma(1/2) = \sqrt{\pi}$.

For reference, we note also the following formulas which can be obtained from (73) by replacing x by $-x$ and combining with the original formulas:

$$\int_{-\infty}^\infty \frac{e^{ikx}}{|x|^{1/2}}\, dx = \sqrt{2\pi}\,|k|^{-1/2}, \tag{77}$$

$$\int_{-\infty}^\infty \frac{\operatorname{sgn} x}{|x|^{1/2}}\, e^{ikx}\, dx = \sqrt{2\pi}\,i|k|^{-1/2}\operatorname{sgn} k. \tag{78}$$

FINITE PARTS OF INTEGRALS

Consider the function

$$f(x) = x^\alpha, \qquad x > 0$$
$$= 0, \qquad x < 0$$

where $-1 < \alpha < 0$. This function defines a distribution if we assign to every testing function $\varphi(x)$ the value of the integral $\int_0^\infty x^\alpha \varphi(x)\, dx$. We shall denote the symbolic function of this distribution by the symbol $x^\alpha H(x)$. To find the derivative of $x^\alpha H(x)$ with respect to x (indicated by an apostrophe), we use

$$\int [x^\alpha H(x)]' \, \varphi(x)\, dx \;=\; -\int x^\alpha H(x)\, \varphi'(x)\, dx \;=\; -\int_0^\infty x^\alpha \varphi'(x)\, dx \tag{79}$$

At this point, we should like to use integration by parts to express the right-hand side of (79) as an integral containing $\varphi(x)$. However, integration by parts would produce an integral of the form $\alpha \int_0^\infty x^{\alpha-1} \varphi(x)\, dx$, and this integral would be divergent at $x = 0$ (since $\alpha < 0$). Nevertheless, we shall write

$$\alpha \int_0^\infty x^{\alpha-1}\, \varphi(x)\, dx \;=\; -\int_0^\infty x^\alpha \varphi'(x)\, dx , \tag{80}$$

and we shall mean by this equation that the meaningless left-hand side (because it is divergent) is defined by the right-hand side; in other words, $x^{\alpha-1}$ is a symbolic function.

A transformation of the right-hand side of (80) will give an apparently different definition for the left-hand side. We have

$$-\int_0^\infty x^\alpha \varphi'(x)\, dx \;=\; -\lim_{\epsilon \to 0} \int_\epsilon^\infty x^\alpha \varphi'(x)\, dx \;=\; \lim_{\epsilon \to 0} \left[\int_\epsilon^\infty \alpha x^{\alpha-1} \varphi(x)\, dx + \epsilon^\alpha \varphi(\epsilon) \right];$$

therefore

$$\int [x^\alpha H(x)]' \, \varphi(x)\, dx \;=\; \int_0^\infty \alpha x^{\alpha-1}\, \varphi(x)\, dx$$

$$\;=\; \lim_{\epsilon \to 0} \left[\int_\epsilon^\infty \alpha x^{\alpha-1}\, \varphi(x)\, dx + \epsilon^\alpha \varphi(\epsilon) \right]. \tag{81}$$

The third member of this equation was first considered by Hadamard, who called it the *finite part* of the divergent integral in the second member.

The idea of the finite part of a divergent integral may be extended to more general integrals of the form $\int_a^b g(x)\, dx$, where $g(x)$ is infinite at $x = a$. Suppose $g(x) = (x-a)^{\alpha-n} h(x)$, where $-1 < \alpha < 0$, n is an integer, and $h(x)$ is nonzero and analytic at $x = a$. We write

$$h(x) \;=\; h(a) + h'(a)(x-a) + \frac{h''(a)}{2!} (x-a)^2 + \ldots + \frac{h^{(n-1)}(a)}{(n-1)!} (x-a)^{n-1}$$

$$+ \,(x-a)^n h_n(x) ,$$

where $h_n(x)$ is again an analytic function of x. If we use this expression for $h(x)$ in the integral, we get

$$\int_{a+\epsilon}^{b} g\,(x)\,dx = \sum_{0}^{n-1} \frac{h^{(k)}(a)}{k!} \int_{a+\epsilon}^{b} (x-a)^{\alpha+k-n}\,dx + \int_{a+\epsilon}^{b} (x-a)^{\alpha} h_n(x)\,dx\,.$$

$$= \sum_{0}^{n} \frac{h^{(k)}(a)}{k!} \left[\frac{(b-a)^{\alpha+k+1-n}}{\alpha+k+1+n} - \frac{\epsilon^{\alpha+k+1-n}}{\alpha+k+1-n} \right]$$

$$+ \int_{a+\epsilon}^{b} (x-a)^{\alpha} h_n(x)\,dx\,.$$

As ϵ approaches zero, the ϵ terms in the bracket become infinite. If we neglect these terms, the remaining terms are finite. We define the finite part (F.P.) of the integral as the sum of the remaining terms, that is,

$$\text{F.P.} \int_{a}^{b} g\,(x)\,dx = \sum_{0}^{n-1} \frac{h^{(k)}(a)}{k!} \frac{(b-a)^{\alpha+k+1-n}}{\alpha+k+1-n} + \int_{a}^{b} (x-a)^{\alpha} h_n(x)\,dx\,. \qquad (82)$$

Hereafter, whenever we write divergent integrals of this form, we shall mean the finite parts of such integrals.

From (81) it is natural to write

$$[x^{\alpha} H(x)]' = \alpha x^{\alpha-1} H(x)\,, \qquad (83)$$

where the right-hand side is to be understood as the symbolic function which produces the finite part of the integral $\int_{0}^{\infty} \alpha x^{\alpha-1} \varphi(x)\,dx$. Equation (83) has so far been proved only for $-1 < \alpha < 0$. It is trivial that it is valid also for $\alpha > 0$. Use of the definition (82) will show that it is valid also for negative values of α, except where α is a negative integer.

The difficulty at negative integral values arises in the definition of the finite part of the integral $\int_{a}^{b} g(x)\,dx$, where $g(x) = (x-a)^{-n-1} h(x)$ and $h(x)$ is analytic at $x = a$. We again write

$$h(x) = \sum_{0}^{n} \frac{h^{(k)}(a)}{k!} (x-a)^{k} + (x-a)^{n+1} h_n(x)\,.$$

We find that

$$\int_{a+\epsilon}^{b} g\,(x)\,dx = \sum_{0}^{n-1} \frac{h^{(k)}(a)}{k!} \left[\frac{(b-a)^{k-n}}{k-n} - \frac{\epsilon^{k-n}}{k-n} \right]$$

$$+ \frac{h^{(n)}(a)}{n!} \left[\log(b-a) - \log\epsilon \right] + \int_{a+\epsilon}^{b} h_n(x)\,dx\,.$$

As ϵ approaches zero, the ϵ terms in the bracket and the $\log \epsilon$ term become infinite. If we neglect these terms, the remaining terms are finite, and we may define the finite part of the integral as

$$\text{F.P.} \int_a^b (x-a)^{-n-1} h(x)\, dx = \sum_0^{n-1} \frac{h^{(k)}(a)}{k!} \frac{(b-a)^{k-n}}{k-n}$$

$$+ \frac{h^{(n)}(a)}{n!} \log(b-a) + \int_a^b h_n(x)\, dx. \tag{84}$$

We may write this result in another, more useful, form as follows:

$$\int_a^b (x-a)^{-n-1} h(x)\, dx = \lim_{\epsilon \to 0} \left[\int_{a+\epsilon}^b (x-a)^{-n-1} h(x)\, dx \right.$$

$$+ \left. \sum_0^{n-1} \frac{h^{(k)}(a)}{k!} \frac{\epsilon^{k-n}}{k-n} + \frac{h^{(n)}(a)}{n!} \log \epsilon \right]. \tag{85}$$

We shall use (85) to find the derivative of the symbolic function $x^{-n} H(x)$. By definition

$$\int \left[x^{-n} H(x) \right]' \varphi(x)\, dx = -\int x^{-n} H(x)\, \varphi'(x)\, dx = -\int_0^\infty x^{-n} \varphi'(x)\, dx$$

$$= -\lim_{\epsilon \to 0} \left[\int_\epsilon^\infty x^{-n} \varphi'(x)\, dx \right.$$

$$+ \left. \sum_0^{n-2} \frac{\varphi^{(k+1)}(0)}{k!} \frac{\epsilon^{k-n+1}}{k-n+1} + \frac{\varphi^{(n)}(0)}{(n-1)!} \log \epsilon \right]$$

by (85). But $\int_\epsilon^\infty x^{-n} \varphi'(x)\, dx = n \int_\epsilon^\infty x^{-n-1} \varphi(x)\, dx - \epsilon^{-n} \varphi(\epsilon)$ by integration by parts; therefore

$$\int [x^{-n} H(x)]' \varphi(x)\, dx = -\lim_{\epsilon \to 0} \left[n \int_\epsilon^\infty x^{-(n+1)} \varphi(x)\, dx - \epsilon^{-n} \varphi(\epsilon) \right.$$

$$+ \left. \sum_0^{n-2} \frac{\varphi^{(k+1)}(0)}{k!} \frac{\epsilon^{k-n+1}}{k-n+1} + \frac{\varphi^{(n)}(0)}{(n-1)!} \log \epsilon \right]. \tag{86}$$

From the Taylor expansion for $\varphi(\epsilon)$, we find that

$$\sum_0^{n-2} \frac{\varphi^{(k+1)}(0)}{k!} \frac{\epsilon^{k-n+1}}{k-n+1} - \epsilon^{-n} \varphi(\epsilon)$$

$$= n \sum_0^{n-1} \frac{\varphi^{(k)}(0)}{k!} \frac{\epsilon^{k-n}}{k-n} - \frac{\varphi^{(n)}(0)}{n!} - \sum_{n+1}^\infty \frac{\varphi^{(k)}(0)}{k!} \epsilon^{k-n}.$$

Substituting this in (86) and using (85), we have

$$\int [x^{-n} H(x)]' \varphi(x)\, dx = -n \int_0^\infty x^{-(n+1)} \varphi(x)\, dx + \frac{\varphi^{(n)}(0)}{n!};$$

therefore, we conclude that

$$[x^{-n} H(x)]' = nx^{-(n+1)} H(x) + \frac{\delta^{(n)}(x)(-1)^n}{n!}. \tag{87}$$

So far, we have defined finite parts of integrals which become infinite at one end point, but it is easy to extend the definition to integrals which have an infinity inside the interval of integration. We put

$$\int (x-a)^{-\alpha-n} h(x)\, dx = \left[\text{F.P.} \int_{-\infty}^a + \text{F.P.} \int_a^\infty \right] (x-a)^{-\alpha-n} h(x)\, dx. \tag{88}$$

With this definition, it is easy to show that $(x^{-a})' = -\alpha x^{-(\alpha+1)}$ for all values of α, integral or not.

In case $\alpha + n = 1$, the definition (88) can be simplified. It becomes

$$\int (x-a)^{-1} h(x)\, dx = \text{P.V.} \int (x-a)^{-1} h(x)\, dx$$

$$= \lim_{\epsilon \to 0} \left[\int_{-\infty}^{a-\epsilon} + \int_{a+\epsilon}^\infty \right] (x-a)^{-1} h(x)\, dx,$$

and is then called the Cauchy *principal value* of the integral on the left-hand side; this elaborates on the last term of (64). For $\alpha + n \neq 1$, the principal value definition cannot be used because it still contains infinite parts.

The formula (58) is an example of a principal value integral. The inversion of formula (72) will be an illustration of a finite part integral. Consider

$$A = \frac{1}{2\pi} \int_{-\infty}^\infty 2 i^{n+1} (n!) \frac{e^{-ikx}}{k^{n+1}}\, dk.$$

To evaluate this integral, we shall use the contour in the k-plane as in Figure 1.3:

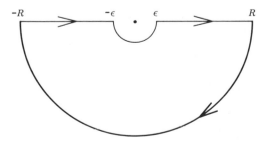

Figure 1.3

Since the integrand of A has no singularity inside this contour, we have the relation

$$A + \int_\epsilon + \int_R = 0, \tag{89}$$

where $\int_\epsilon$ and $\int_R$ denote the limits of the corresponding integrals over semi-circles of radius ϵ and R, respectively. For $x > 0$, the integral over the semicircle of radius R vanishes as R approaches infinity. The integral over the semicircle of radius ϵ becomes infinite as ϵ approaches zero, but if we take the finite part of the integral we find

$$\text{F.P.} \int_\epsilon \frac{e^{-ikx}}{k^{n+1}} dk = \text{F.P.} \ i \int_{-\pi}^0 \exp\left[-i\epsilon x e^{i\theta}\right] e^{-ni\theta} d\theta \ \epsilon^{-n}$$

$$= \text{F.P.} \ i \int_{-\pi}^0 \sum_0^\infty (-ix)^k \frac{e^{i(k-n)\theta}}{k!} d\theta \ \epsilon^{k-n}$$

$$= (-1)^n \ i^{n+1} \frac{(+\pi)}{n!} x^n. \tag{90}$$

From (89) and (90), we get

$$A = -\int_\epsilon = \frac{(-1)^{n+1}}{\pi} \ i^{2(n+1)} \ (n!) \ \frac{\pi}{n!} x^n = x^n,$$

for $x > 0$. If $x < 0$, we take a similar contour in the upper half-plane and we find $A = -x^n$; therefore $A = x^n \ \text{sgn} \ x$, which verifies (72).

THE SURFACE AREA OF THE n-DIMENSIONAL SPHERE

As an application of these formulas, we shall derive the surface area of the n-dimensional sphere of radius r, that is, the surface area of the region R bounded by $x_1^2 + x_2^2 + \ldots + x_{n+1}^2 < r^2$. The volume of the sphere is given by the integral

$$V(r) = \int_R \cdots \int dx_1 \cdots dx_{n+1} = \int \cdots \int H\left(r^2 - \Sigma x_j^2\right) dx_1 \cdots dx_{n+1}, \tag{91}$$

where $H(t)$ denotes the Heaviside unit function, and the second integral is extended over all of $(n + 1)$-dimensional space. If we let $S(r)$ denote the surface area of the sphere, then

$$S(r) = \frac{dV(r)}{dr} = \int \cdots \int 2r\delta(r^2 - \Sigma x_j^2) \ dx_1 \cdots dx_{n+1}. \tag{92}$$

From (51), we know that

$$\delta(r^2 - \Sigma x_j^2) = \frac{1}{2\pi} \int e^{ik(r^2 - \Sigma x_j^2)} dk.$$

Substituting this in (92), we obtain

$$S(r) = \frac{2r}{2\pi} \int e^{ikr^2} dk \left(\int e^{-iky^2} dy\right)^{n+1}. \tag{93}$$

We write

$$\int e^{-iky^2} dy = \lim_{\epsilon \to 0} \int e^{-(\epsilon + ik)y^2} dy = \lim_{\epsilon \to 0} \left(\frac{\pi}{\epsilon + ik}\right)^{1/2}$$

by a well-known formula. Then, with the help of this result, (93) becomes

$$S = \frac{r}{\pi} \lim_{\epsilon \to 0} \int e^{ikr^2} \left(\frac{\pi}{\epsilon + ik}\right)^{\frac{n+1}{2}} dk. \tag{94}$$

The results depend on whether n is odd or even. If n is odd, say equal to $2m + 1$, then

$$S - \pi^{(n-1)/2} r \lim_{\epsilon \to 0} \int \frac{e^{ikr^2} dk}{(\epsilon + ik)^{m+1}}.$$

Considering a contour which is closed by a semicircle in the upper half-plane, we find that S equals the residue at the $(m+1)$th-order pole $k = +i\epsilon$. Thus

$$S = 2 \frac{\pi^{(n+1)/2} r^n}{[(n-1)/2]!}. \tag{95}$$

If n is even, equal to $2m$, then

$$S = \pi^{(n-1)/2} r \int \frac{e^{ikr^2}}{(ik)^{m+1/2}} dk.$$

Using the inverse Fourier transform to formula (76), we obtain

$$S = \frac{2\pi^{(n+1)/2} r^n}{\Gamma\left(\frac{n+1}{2}\right)}. \tag{96}$$

Formulas (95) and (96) are the desired result for the surface area of the sphere.

OPERATIONAL METHODS AND LAPLACE TRANSFORM

We wish to discuss operational methods for solving differential equations with initial conditions. For example, consider the problem of finding the solution of

$$u' + u = 1 \tag{97}$$

such that $u(0) = 1$. Since we are interested in $u(t)$ only for positive values of t, we may assume that $u(t)$ is identically zero for negative values of t. Because of this fact, we may denote the solution of (97) by $v(t) = u(t)H(t)$, where $H(t)$ is again Heaviside's unit function. This notation will be useful because we can now write a differential equation for $v(t)$, which will automatically include the initial condition on $u(t)$.

As we have seen in Section 1.1, the derivative of $v(t)$ will contain a delta function at the origin because of the discontinuity in $v(t)$ at $t = 0$. We find that $v' = u'H(t) + u(0)\delta(t) = u'(t)H(t) + \delta(t)$. Comparing this with (97), we get the following differential equation for $v(t)$: $v' + v = H(t) + \delta(t)$ or, equivalently,

$$(p + 1)v = H(t) + \delta(t) \tag{98}$$

where p denotes the differential operator.

Following Heaviside, we shall assume that p can be treated as if it were an algebraic quantity. Then the solution of (98) is

$$v = \frac{1}{p+1} \left[H(t) + \delta(t) \right];$$

but we know that $\delta(t) = pH(t)$; therefore

$$v = \frac{1}{p+1} (1+p)H(t) = H(t),$$

and consequently, the solution of (97) is $u(t) = 1$.

This simple example indicates the method that can be used to solve any linear differential equation with constant coefficients. Consider the problem of finding the solution of

$$Lu = \frac{d^n u}{dt^n} + a_1 \frac{d^{n-1}u}{dt^{n-1}} + \ldots + a_{n-1}\frac{du}{dt} + a_n u = f(t), \tag{99}$$

such that

$$u(0) = C_0, \ u'(0) = C_1, \ldots, u^{(n-1)}(0) = C_{n-1}. \tag{100}$$

We proceed as before. Let $v(t) = u(t)H(t)$, then

$$Lv = (p^n + a_1 p^{n-1} + \ldots + a_n)v$$

$$= f(t)H(t) + C_0'\delta^{(n-1)}(t) + C_1'\delta^{(n-2)}(t) + \ldots + C_{n-1}'\delta(t)$$

$$= f(t)H(t) + (C_0'p^{n-1} + C_1'p^{n-2} + \ldots + C_{n-1}')\delta(t) \tag{101}$$

where

$$C_0' = C_0, \qquad C_1' = C_1 + a_1 C_0, \quad \dots ,$$

$$C_{n-1}' = C_{n-1} + a_1 C_{n-2} + \dots + a_{n-2} C_1 + a_{n-1} C_0 .$$

The solution of this equation can be written as

$$v = (p^n + a_1 p^{n-1} + a_1 p^{n-1} + \dots + a_n)^{-1}$$

$$[f(t) H(t) + (C_0' p^{n-1} + C_1' p^{n-2} + \dots + C_{n-1}') \delta(t)], \qquad (102)$$

but this result must still be interpreted.

In (102) there are two kinds of terms that need interpretation. One is $\varphi(p)\delta(t)$, where $\varphi(p)$ is a rational function of p; the other is $\psi(p) f(t) H(t)$ where $\psi(p)$ is also a rational function of p. However, the second type of term can be reduced to the first because $\psi(p) f(t) H(t) = \psi(p) \int f(\tau) H(\tau) \delta(t-\tau) d\tau = \int f(\tau) H(\tau) \psi(p) \delta(t-\tau) dt$, and we see that what is needed is an interpretation for $\psi(p)\delta(t-\tau)$, a term of the first type.

Before interpreting terms of the first type, we shall consider some simple examples of differentiation which will be useful. We shall always use p to mean differentiation with respect to t. It is easy to verify that

$$p^{n+1} t^n H(t) = n! \, \delta(t), \qquad (p - a) e^{at} H(t) = \delta(t).$$

These results can be rewritten as follows:

$$p^{-(n+1)} \delta(t) = \frac{t^n H(t)}{n!}, \qquad (103)$$

$$(p - a)^{-1} \delta(t) = e^{at} H(t), \qquad (104)$$

and we thus have the interpretation of some simple operators on $\delta(t)$. Another useful result is obtained by differentiating (104) n times with respect to a. We get

$$(p - a)^{-(n+1)} \delta(t) = \frac{t^n e^{at}}{n!} H(t). \qquad (105)$$

These formulas (103)-(105) are sufficient to interpret (102). Consider a term such as $\varphi(p)\delta(t)$, where $\varphi(p)$ is a rational function of p; for simplicity we assume that $\varphi(p)$ has only simple factors. By well-known methods $\varphi(p)$ may be expressed in terms of partial fractions as

$$\varphi(p) = \sum \frac{\alpha_k}{p - a_k},$$

and then by (104),

$$\varphi(p)\ \delta(t) = \sum \frac{\alpha_k}{p - a_k}\ \delta(t) = \sum \alpha_k \exp(a_k t)\ H(t)\ . \tag{106}$$

There remains the general question: what is meant by $\varphi(p)\delta(t)$, where $\varphi(p)$ need not be rational? An answer to this question may be given by using formula (51). We have

$$\delta(t) = \frac{1}{2\pi} \int_{-\infty}^{\infty} e^{ikt}\, dk\ .$$

If we put $ik = s$, then

$$\delta(t) = \frac{1}{2\pi i} \int_{-i\infty}^{i\infty} e^{st}\, ds\ ,$$

and

$$\varphi(p)\delta(t) = \frac{1}{2\pi i} \int_{-i\infty}^{i\infty} \varphi(p)\, e^{st}\, ds\ ;$$

but it is easy to verify that, if $\varphi(p)$ is an analytic function of p, $\varphi(p)\, e^{st} = \varphi(s)\, e^{st}$ and therefore

$$\varphi(p)\delta(t) = \frac{1}{2\pi i} \int_{-i\infty}^{i\infty} \varphi(s)\, e^{st}\, ds\ . \tag{107}$$

The integral in (107) is known as *Bromwich's integral*. However, the path of integration is usually taken along a line in the right half-plane parallel to the imaginary axis and not exactly on the imaginary axis. Since the integral will usually vanish for negative values of t, we may represent its value by $F(t)H(t)$. Equation (107) may then be written

$$\varphi(p)\delta(t) = F(t)H(t)\ . \tag{108}$$

We call $\varphi(p)$ the *Laplace transform* of $F(t)$. Simple examples of Laplace transforms are given in (103) - (105), namely, the Laplace transforms of t^n, e^{at}, and $t^n e^{at}$, are at $n!\,p^{-(n+1)}$, $(p - a)^{-1}$, and $n!\,(p - a)^{-(n+1)}$, respectively.

SOLUTION OF THE HEAT EQUATION

Consider the problem of finding the solution of the heat equation

$$u_{xx} = u_t \tag{109}$$

which satisfies the condition

$$u(x, 0) = 0\ ,\quad u_x(0, t) = f(t)\ . \tag{110}$$

Putting $u(x, t)H(t) = v(x, t)$ and using p as before to denote differentiation with respect to t, we get the following equation for v:

$$v_{xx} = pv \tag{111}$$

with $v_x = f(t)$ for $x = 0$. The solution of (111) satisfying this boundary condition is

$$v = -\frac{e^{-xp^{\frac{1}{2}}}}{p^{1/2}} f(t)H(t).$$ (112)

To interpret this formula, it is necessary to know the meaning of

$$w = -\frac{e^{-xp^{\frac{1}{2}}}}{p^{1/2}} \delta(t).$$

We shall find this meaning by a trick, but first we need the following. *Lemma:* Let $a(p)$ be a function of p and $a'(p)$ its derivative with respect to p; then

$$a(p)t\,\varphi(t) = t\,a(p)\varphi(t) + a'(p)\varphi(t).$$ (113)

The proof of this lemma will follow from the well-known rule for the derivative of a product, namely, $pu(t)v(t) = u(t)pv(t) + v(t)pu(t)$. We write this result in a more useful form by putting $p = p_1 + p_2$, where p_1 will act only on $u(t)$ and p_2 will act only on $v(t)$, as follows:

$$pu(t)v(t) = (p_1 + p_2)u(t)v(t).$$

In this form it is clear that $a(p)u(t)v(t) = a(p_1 + p_2)u(t)v(t)$. Let $u(t) = t$ and $v(t) = \varphi(t)$ and apply Taylor's theorem to $a(p_1 + p_2)$. We have $a(p_2 + p_1) = a(p_2) + p_1 a'(p_2) + \ldots$, so that

$$a(p)t\,\varphi(t) = t\,a(p_2)\varphi(t) + p_1 t\,a'(p_2)\varphi(t) + 0,$$ (114)

since $p_1^n t = 0$ for $n > 1$. Since $p_1 t = 1$, we see that (114) is the same as (113).

We now return to the evaluation of w. Consider tw_x. We find

$$tw_x = te^{-xp^{\frac{1}{2}}}\delta(t) = e^{-xp^{\frac{1}{2}}}t\delta(t) - \frac{x}{2}w$$

by the use of the lemma; since $t\,\delta(t) = 0$, this implies that $tw_x = -(x/2)w$. Integrating, we get $w = C\exp(-x^2/4t)$, where C is independent x. From the definition, $w = -p^{-1/2}\delta(t)$ for $x = 0$, and from Bromwich's integral

$$\frac{-1}{2\pi i}\int e^{pt}p^{-1/2}\,dp = (\pi t)^{-1/2}H(t),$$

we conclude that

$$w = (\pi t)^{-1/2}\exp\left(-\frac{x^2}{4t}\right)H(t).$$ (115)

With this result we can evaluate v. We find

$$v = -\frac{e^{-xp^{\frac{1}{2}}}}{p^{1/2}} f(t) H(t) = -\frac{e^{-xp^{\frac{1}{2}}}}{p^{1/2}} \int f(\tau) H(\tau) \delta(t - \tau) d\tau \tag{116}$$

$$= \int f(\tau) H(\tau) \left[-\frac{e^{-xp^{\frac{1}{2}}}}{p^{1/2}} \delta(t - \tau) \right] d\tau$$

$$= \int f(\tau) H(\tau) \left[\pi^{-1/2} (t - \tau)^{-1/2} \exp\left[\frac{-x^2}{4(t - \tau)} \right] H(t - \tau) \right] d\tau$$

$$= \pi^{-1/2} \int_0^t (t - \tau)^{-1/2} f(\tau) \exp\left[\frac{-x^2}{4(t - \tau)} \right] d\tau .$$

THE SHIFT THEOREM AND THE CONVOLUTION THEOREM

Let us apply Bromwich's integral (107) to the function e^{-ph}. We find that

$$e^{-ph} \delta(t) = \frac{1}{2\pi i} \int_{-i\infty}^{i\infty} e^{s(t-h)} ds = \delta(t-h) \tag{117}$$

by the use of formula (107) for the case where $\varphi(p) = 1$. With the help of (117) we shall now show that the relation between $\varphi(p)$ and $F(t)$ in (108) is actually the Laplace transformation. By the fundamental property of the δ-function, we know that $F(t)H(t) = \int F(\tau)H(\tau) \delta(t - \tau) d\tau$, and by (117) we have

$$F(t)H(t) = \int F(\tau)H(\tau) e^{-p\tau} \delta(t) d\tau = \int_0^\infty F(\tau) e^{-p\tau} d\tau \delta(t) = \varphi(p)\delta(t),$$

the form in (108). We see then that

$$\varphi(p) = \int_0^\infty F(\tau) e^{-p\tau} d\tau \tag{118}$$

is the Laplace transform of $F(t)$.

Formula (117) is a special case of the *Shift theorem* which reads as follows:

$$e^{-ph} f(t) = f(t-h) . \tag{119}$$

The proof results from the use of (117). We have

$$e^{-ph} f(t) = e^{-ph} \int f(\tau) \delta(t - \tau) d\tau = \int f(\tau) \delta(t - h - \tau) d\tau = f(t-h) .$$

Another important theorem is the *Convolution theorem* which enables us to invert the product of two Laplace transforms. This theorem states that if $\varphi(p)$ is the Laplace transform of $F(t)$ and if $\psi(p)$ is the Laplace transform of $G(t)$, then $\varphi(p)\psi(p)$ is the Laplace transform of the convolution of $F(t)$ and $G(t)$, that is, of the integral

$$\int_0^t G(\tau)F(t-\tau)d\tau.$$

Suppose that $K(t)$ is the Laplace transform of $\varphi(p)\psi(p)$. By (108), we have $\varphi(p)\psi(p)\delta(t) = K(t)H(t)$, $\varphi(p)\delta(t) = F(t)H(t)$, $\psi(p)\delta(t) = G(t)H(t)$; therefore

$$\varphi(p)\psi(p)\delta(t) = \varphi(p)G(t)H(t) = \varphi(p)\int G(\tau)H(\tau)\delta(t-\tau)d\tau$$

$$= \int G(\tau)H(\tau)\varphi(p)\delta(t-\tau)d\tau$$

$$= \int G(\tau)H(\tau)F(t-\tau)H(t-\tau)d\tau = \int_0^t G(\tau)F(t-\tau)d\tau.$$

This proves the theorem.

FOURIER SERIES

The methods of the Theory of Distributions can also be applied to the study of Fourier series. We begin with a discussion of the series for the logarithm. By Taylor's theorem we have

$$-\log(1-z) = +z + \frac{z^2}{2} + \frac{z^3}{3} + \dots \tag{121}$$

Put $z = re^{i\theta}$ and note that $\log(1-re^{i\theta}) = (1/2)\log(1-2r\cos\theta + r^2) - i\tan^{-1}[r\sin\theta/(1-r\cos\theta)]$. Using this in (121) and separating real and imaginary parts, we get

$$r\cos\theta + \frac{r^2}{2}\cos 2\theta + \frac{r^3}{3}\cos 3\theta + \dots = -\frac{1}{2}\log(1-2r\cos\theta+r^2), \tag{122}$$

$$r\sin\theta + \frac{r^2}{2}\sin 2\theta + \frac{r^3}{3}\sin 3\theta + \dots = \tan^{-1}\frac{r\sin\theta}{1-r\cos\theta}. \tag{123}$$

Put $r = 1$ in these results, note that $1 - \cos\theta = 2\sin^2(\theta/2)$, $\sin\theta = 2\cos(\theta/2)\sin(\theta/2)$, and we have

$$\cos\theta + \frac{\cos 2\theta}{2} + \frac{\cos 3\theta}{3} + \dots = -\log\left(2\sin\frac{\theta}{2}\right), \tag{124}$$

$$\sin\theta + \frac{\sin 2\theta}{2} + \frac{\sin 3\theta}{3} + \dots = \tan^{-1}\left(\cot\frac{\theta}{2}\right), \tag{125}$$

where that branch of $\tan^{-1}$ must be taken which is between $-\pi/2$ and $\pi/2$.

The right-hand side of (125) can be simplified. Since

$$\cot \frac{\theta}{2} = \tan\left(\frac{\pi}{2} - \frac{\theta}{2}\right),$$

we have

$$\tan^{-1}\left(\cot\frac{\theta}{2}\right) = \frac{\pi - \theta}{2} + n\pi, \tag{126}$$

where n is any integer. Because of the restriction on the branch of $\tan^{-1}$, (126) is valid for

$$2n\pi < \theta < 2(n+1)\pi. \tag{127}$$

We have, therefore,

$$\sin\theta + \frac{\sin 2\theta}{2} + \frac{\sin 3\theta}{3} + \ldots = \frac{\pi - \theta}{2} + n\pi, \tag{128}$$

if (127) is satisfied.

   Both (124) and (128) are trigonometrical series which converge to the right-hand sides in the ordinary sense. However, if we differentiate (124) and (128) with respect to θ, we shall get series which converge, not in the ordinary sense, but in the distribution sense to certain symbolic functions. In this way we get

$$\sin\theta + \sin 2\theta + \sin 3\theta + \ldots = \frac{1}{2}\cot\frac{\theta}{2}, \tag{129}$$

$$\cos\theta + \cos 2\theta + \cos 3\theta + \ldots = -\frac{1}{2} + \pi\delta_p(\theta), \tag{130}$$

where $\delta_p(\theta)$ denotes the periodic δ-function, that is, a δ-function with singularities at the points $\theta = 0$, $\theta = \pm 2n\pi$, where n is any integer. This term in (130) is a consequence of the fact that (128) is valid only under the restriction (127).

   We may rewrite (130) in the following more useful form:

$$\frac{1}{2} + \sum_{1}^{\infty} \cos k\theta = \pi \sum_{-\infty}^{\infty} \delta(\theta - 2n\pi).$$

Replace θ by $\theta - \theta'$ and we get the fundamental formula for Fourier series, namely,

$$\frac{1}{2} + \sum_{1}^{\infty} \cos k(\theta - \theta') = \pi \sum_{-\infty}^{\infty} \delta(\theta - \theta' - 2n\pi). \tag{131}$$

To see the relation between (131) and Fourier series, multiply it by a function $f(\theta')$ which is defined for $-\pi < \theta' < \pi$ and integrate between $-\pi$ and π. We get

$$\frac{1}{2\pi} \int_{-\pi}^{\pi} f(\theta') \, d\theta' + \frac{1}{\pi} \int_{-\pi}^{\pi} \sum f(\theta') \cos k(\theta - \theta') \, d\theta' = f(\theta),$$

or

$$\frac{1}{2\pi} \int_{-\pi}^{\pi} f(\theta') \, d\theta' + \sum \cos k\theta \, \frac{1}{\pi} \int_{-\pi}^{\pi} f(\theta) \cos k\theta' \, d\theta' \qquad (132)$$

$$+ \sum \sin k\theta \frac{1}{\pi} \int_{-\pi}^{\pi} f(\theta') \sin k\theta' \, d\theta' = f(\theta), \qquad -\pi < \theta < \pi.$$

This is, of course, the well-known formula for the Fourier series for $f(\theta)$. There is another interesting formula that may be obtained from (131). Replace the cosines by their expression in terms of exponentials, and (131) becomes

$$\sum_{-\infty}^{\infty} e^{ik(\theta - \theta')} = 2\pi \sum_{-\infty}^{\infty} \delta(\theta - \theta' - 2n\pi). \qquad (133)$$

Let $\varphi(\theta)$ be a testing function and let $F(k)$ be its Fourier transform so that

$$F(k) = \int_{-\infty}^{\infty} e^{ik\theta} \varphi(\theta) \, d\theta.$$

Put $\theta' = 0$ in (133), multiply by $\varphi(\theta)$ and integrate from $-\infty$ to ∞. We get *Poisson's formula,*

$$\sum_{-\infty}^{\infty} F(k) = 2\pi \sum_{-\infty}^{\infty} \varphi(2n\pi). \qquad (134)$$

A TRAPEZOIDAL PULSE

Since we have shown in Section 1.2 that if a sequence s_n of symbolic functions converges to a symbolic function, then the sequence of derivatives s_n' converges to the derivative s', it follows that a Fourier series may be differentiated term by term as frequently as we wish. This fact enables us to find easily the Fourier series of certain complicated functions. We shall illustrate the procedure by finding the Fourier series for the periodic trapezoidal pulse with period 2π shown in Figure 1.4.

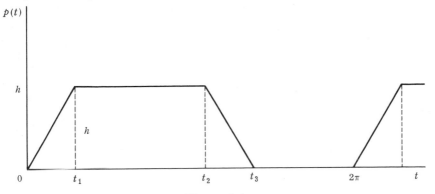

Figure 1.4

Suppose that $p(t) = \sum_{-\infty}^{\infty} a_k e^{ikt}$, then $p'(t) = \sum ika_k e^{ikt}$ and $p''(t) = -\sum k^2 a_k e^{ikt}$. We have

$$p'(t) = h/t_1, \qquad\qquad 0 < t < t_1$$

$$= 0, \qquad\qquad t_1 < t < t_2$$

$$= -h(t_3 - t_2)^{-1}, \qquad t_2 < t < t_3$$

$$= 0, \qquad\qquad t_3 < t < 2\pi,$$

and

$$p''(t) = (h/t_1)\delta(t) - (h/t_1)\delta(t - t_1) - h(t_3 - t_2)^{-1}\delta(t - t_2)$$

$$+ h(t_3 - t_2)^{-1}\delta(t - t_3).$$

By (133)

$$\delta(t - t') = \frac{1}{2\pi}\sum_{-\infty}^{\infty} e^{ik(t-t)};$$

therefore,

$$p''(t) = -\sum k^2 a_k e^{ikt} = \frac{1}{2\pi}\sum e^{ikt}\left[\frac{h}{t_1}\left(1 - e^{-ikt_1}\right) - \frac{h}{t_3 - t_2}\left(e^{-ikt_2} - e^{-ikt_3}\right)\right].$$

This shows that

$$-k^2 2\pi a_k = \frac{h}{t_1}\left(1 - e^{-ikt_1}\right) - \frac{h}{t_3 - t_2}\left(e^{-ikt_2} - e^{-ikt_3}\right), \tag{135}$$

and thus defines a_k for all values of k except $k = 0$. To obtain a_0 we notice that

$$a_0 = \frac{1}{2\pi}\int_{-\pi}^{\pi} p(t)\,dt = \frac{1}{2\pi}\left[h\frac{(t_1 + t_3 - t_2)}{2} + h(t_2 - t_1)\right]. \tag{136}$$

Combining (135) and (136), we obtain finally

$$p(t) = \frac{1}{2\pi}h\frac{(t_3 + t_2 - t_1)}{2} - h\sum\left[\frac{1 - e^{-ikt_1}}{t_1} - \frac{e^{-ikt_2} - e^{-ikt_3}}{t_3 - t_2}\right]\frac{e^{ikt}}{2\pi k^2}.$$

SUMMING A FOURIER SERIES

Symbolic differentiation of Fourier series may also help in summing the series. The idea is to differentiate the series and to combine the resulting series in such a way that a series whose sum is known is obtained. For example, consider the series

$$S = \frac{1}{2a^2} + \sum_1^\infty \frac{\cos nx}{a^2 + n^2}. \tag{137}$$

Differentiating this series with respect to x, we obtain

$$S' = -\sum_1^\infty \frac{n \sin nx}{a^2 + n^2}, \tag{138}$$

and

$$S'' = -\sum \frac{n^2 \cos nx}{a^2 + n^2}. $$

Combining this last result with a^2 times the original series (137), we get

$$S'' - a^2 S = -\frac{1}{2} - \sum_1^\infty \cos nx = -\pi\, \delta\,(x) \tag{139}$$

by (131).

The problem of summing the series is thus equivalent to finding a solution of the differential equation

$$S'' - a^2 S = -\pi\, \delta\,(x). \tag{140}$$

This solution will not be unique unless we adjoin some boundary conditions. From (138) we see that $S'(\pi) = S'(-\pi) = 0$. We shall now find a solution of (140) satisfying these conditions.

The solutions of the homogeneous equation corresponding to (140) are $e^{\pm ax}$. A suitable combination of these that satisfies the condition of vanishing at $x = \pi$ is $\sinh a\,(\pi - x)$ and that satisfies the condition of vanishing at $x = -\pi$ is $\sinh a\,(\pi + x)$. Thus the boundary conditions on S' give

$$S = A \cosh a\,(\pi + x), \qquad -\pi < x < 0,$$

$$= B \cosh a\,(\pi - x), \qquad 0 < x < \pi,$$

where the constants A and B must be so chosen that $S\,(x)$ is continuous at $x = 0$ and $S'\,(x)$ has a jump of magnitude $-\pi$ at $x = 0$. These conditions imply that

$$A = B,$$

$$(A + B)a \sinh a\pi = \pi;$$

therefore,

$$S = \frac{\pi \cosh a(\pi + x)}{2a \sinh a\pi}, \qquad -\pi < x < 0;$$

$$S = \frac{\pi \cosh a(\pi - x)}{2a \sinh a\pi}, \qquad 0 < x < \pi. \tag{141}$$

If we put $x = 0$ in both (137) and (141), we get the well-known expansion for $\coth \pi a$, namely,

$$\frac{\pi}{2a} \coth a\pi = \frac{1}{2a^2} + \sum_{1}^{\infty} \frac{1}{n^2 + a^2} \tag{142}$$

Near $a = 0$, we have

$$\frac{\pi}{2a} \coth a\pi = \frac{1}{2a^2} + \frac{\pi^2}{6} + O(a^2).$$

Substituting this in (142) and letting a approach zero, we get $\sum_{1}^{\infty}(1/n^2) = (\pi^2/6)$, a familiar result.

Spectral Theory of Operators

2.1 LINEAR VECTOR SPACES

INTRODUCTION

The theory of linear vector spaces is a generalization of the theory of three-dimensional vectors. By means of this general theory we can apply geometric insight and intuition to complicated problems in analysis. Of course, the geometric approach is no substitute for an analytic proof; however, by analogy with corresponding problems in two or three dimensions, the geometric approach does suggest possible methods for analysis.

To explain the concept of linear vector spaces, we shall start first with the definition of a "space." To a mathematician, a *space* is a collection of objects which are called the *elements* of the space. The elements might be numbers, points, functions, matrices, events, or anything whatsoever. The concept "space" is almost synonymous with that of "class" or "set." Usually, however, the term "space" is more inclusive: it is essentially equivalent to the logician's or philosopher's term "Universe of Discourse." For example, the mathematician would call the collection of all points on a given straight line a space; a collection of points on this line would be called a set, and a collection of sets would be called a class.

It is clear that because this concept of space is too general, spaces as such have few interesting properties. Usually we talk about special kinds of spaces which have special properties. The properties we consider are specified by the name of the space. For example, a topological space is a space in which a topology, that is, the notion of closeness of points, is defined; a metric space is a space in which a metric, that is, the distance between two points, is defined.

A *linear vector space* is a space whose elements are called vectors in which the notions of addition of two vectors and multiplication of a vector by a real or complex number are defined. To give a more precise definition, consider a space S with elements which we denote by $x, y, z, \dots$. We say that S is a linear vector space if we can define a notion of addition so that $x + y$ is an element in S, and if we can define a notion of scalar multiplication so that αx, where α is a number, is an element in S. The addition operation should have the following properties:

(a) $(x + y) + z = x + (y + z)$;

(b) $x + y = y + x$;

(c) S contains an element, which we denote by 0, such that for any element x in S, we have $x + 0 = 0 + x = x$;

(d) To every element x in S, we may associate an element, which we denote by $(-x)$, such that $x + (-x) = 0$.

Also, it should be possible to define multiplication by a scalar, such as any real or complex number. For simplicity, we shall at present restrict ourselves to multiplication by real numbers. Let α, β denote arbitrary real numbers, then for any x in S, αx should be an element of S possessing the following properties:

(e) $\alpha(x + y) = \alpha x + \alpha y$;

(f) $(\alpha + \beta)x = \alpha x + \beta x$;

(g) $\alpha(\beta x) = (\alpha\beta)x$;

(h) $1x = x$.

We shall call the elements of S *vectors* and the numbers we multiply them with *scalars*. The result of $x + y$ will be called the *sum* of the two vectors.

EXAMPLES

(1) The space of three-dimensional vectors. The elements are the vectors from the origin O to any point P in three-dimensional space. We denote this vector by the symbol OP. The sum of the vectors OP and OQ is the vector OR which is obtained by the usual parallelogram rule. When OP is multiplied by a scalar α, the result is a vector whose length is $|\alpha|$ times the length of OP and whose direction is the same or opposite to that of OP according as α is positive or negative.

(2) The space of all triplets of real numbers. Here, an element x denotes the three real numbers (ξ_1, ξ_2, ξ_3). If y denotes the triplet (η_1, η_2, η_3), then we define $x + y$ as the element whose triplet is $(\xi_1 + \eta_1, \xi_2 + \eta_2, \xi_3 + \eta_3)$, and αx as the element whose triplet is $(\alpha\xi_1, \alpha\xi_2, \alpha\xi_3)$. It is easy to show that these definitions satisfy properties (a) - (h); therefore, this space is a linear vector space.

The space of Example 2 is essentially the same as the space considered in Example 1. (Mathematically, they are *isomorphic*.) To see this, take any geometric vector OP and consider its components on any three lines, mutually perpendicular or not, passing through O and not all lying in the same plane. These three numbers, the components of OP, are an element of the space defined in Example 2. To every vector in Example 1, there corresponds a unique triplet in Example 2; and conversely, to every triplet in Example 2, there corresponds a unique vector in Example 1, namely, the vector which has these three numbers as its components. We say that there is a *one-to-one correspondence* between the space defined in Example 1 and the space defined in Example 2. This correspondence also extends to the addition of vectors and to multiplication by a scalar; that is, if OP corresponds to a triplet x and OQ corresponds to a triplet y, then $OP + OQ$ corresponds to the triplet $x + y$ and the vector αOP corresponds to the triplet αx. Because of this, any property of the vectors in Example 1 is exactly duplicated by the triplets in Example 2, and conversely. That is why we say the spaces are isomorphic to each other.

However, notice that the plus sign means different things in Examples 1 and 2. In Example 1, plus means geometric addition by the parallelogram rule, whereas in Example 2, plus means ordinary addition of each number of one triplet with the corresponding number of the other triplet.

Notice also that the one-to-one correspondence between the two spaces is not unique because if we were to choose a different set of three lines through O, the components of OP on the new lines would be a different triplet of real numbers. Therefore, to every vector OP, there can be many different representations as a triplet of real numbers.

(3) The space E_n of n-tuplets of real numbers, when $n = 1, 2, 3, \dots$. Here, an element x denotes the n real numbers $(\xi_1, \xi_2, \dots, \xi_n)$. If y denotes the n real numbers $(\eta_1, \eta_2, \dots, \eta_n)$ then we define

$$x + y = (\xi_1 + \eta_1, \xi_2 + \eta_2, \dots, \xi_n + \eta_n),$$

$$\alpha x = (\alpha \xi_1, \alpha \xi_2, \dots, \alpha \xi_n).$$

Again, it is easy to show that these definitions satisfy properties (a) - (h). Note that the space defined in Example 2 is just E_3.

(4) The space E_∞ of all infinite sequences of real numbers. We put

$$x = (\xi_1, \xi_2, \dots),$$

$$y = (\eta_1, \eta_2, \dots),$$

and then if we define

$$x + y = (\xi_1 + \eta_1, \xi_2 + \eta_2, \dots),$$

$$\alpha x = (\alpha \xi_1, \alpha \xi_2, \dots),$$

it is easy to see that E_∞ is a linear vector space.

(5) Function spaces. These will be defined in the next section.

FUNCTION SPACES

Consider the space of all points on a plane or in three-dimensional space. These spaces as such are *not* linear vector spaces. However, a linear vector space can be assigned to them. For example, pick a point O in the space, call it the origin, and to every point P assign the geometric vector OP. As we have mentioned before, these geometric vectors form a linear vector space.

This idea of geometric vectors is restricted to one-, two-, and three-dimensional space and cannot be applied in a general space. However, a linear vector space can be attached to any space S by the following procedure.

Let s denote an element of S. Suppose that $f(s)$ is a real-valued function of s, that is, suppose that to every element s of S a real number, which we denote by $f(s)$, is assigned. We shall say that the function of $f(s)$ defines a *mapping* f of the space S into the space of real numbers E_1. Let g be a mapping which assigns to S the real number $g(s)$; then we may define $f + g$ as the mapping which assigns to s the real number $f(s) + g(s)$, and if α is a real number we may define αf as the mapping which assigns to s the real number $\alpha f(s)$. In this way, the space of all mappings f, or equivalently, the *function space f(s)*, becomes a linear vector space over S.

It is clear that we could do similar things for the mappings of S into E_n, that is, the assignment of n real numbers to every element s. If we let $f_1(s), f_2(s), \ldots, f_n(s)$ denote the n real numbers assigned to the element s by a mapping f and, similarly, let $g_1(s), g_2(s), \ldots, g_n(s)$ denote the n real numbers assigned to the element s by a mapping, then we may define αf as the mapping which assigns the numbers $\alpha f_1(s), \alpha f_2(s), \ldots, \alpha f_n(s)$ to the element s, and we may define $f + g$ as the mapping which assigns the numbers $f_1(s) + g_1(s), f_2(s) + g_2(s), \ldots, f_n(s) + g_n(s)$ to the element s. Again in this way the space of all mappings f, or equivalently, the space of n-tuple functions $f_1(s), f_2(s), \ldots, f_n(s)$, becomes a linear vector space over S.

In many applications it is useful to consider not all mappings f or all functions $f(s)$ but only a certain class of functions. For example, if S is the set of all real numbers s from $-\infty$ to ∞, then let us consider the following classes of mappings or functions:

C, the class of continuous functions on S;

M, the class of non-decreasing functions on S;

L_2, the class of square-integrable functions on S.

It is clear that if $f(s)$ and $g(s)$ belong to C, that is, if $f(s)$ and $g(s)$ are both continuous functions of s, then $\alpha f(s)$ and $f(s) + g(s)$ also belong to C; consequently, C is a linear vector space. However, if $f(s)$ and $g(s)$ belong to M, it is true that $f(s) + g(s)$ belongs to M but, if α is negative, then $\alpha f(s)$ will be nonincreasing and therefore not in M; consequently, M is not a linear vector space. It will be seen later that if $f(s)$ and $g(s)$ belong to L_2, then $\alpha f(s) + \beta g(s)$ will also belong to L_2. This will imply that L_2 is a linear vector space.

In a certain sense the linear vector spaces E_n and E_∞ defined previously may be considered as function spaces. If we put $\xi_1 = f(1)$, $\xi_2 = f(2), \ldots, \xi_n = f(n)$, we see that E_n is the space of all functions defined on the n integers $1, 2, \ldots, n$. Similarly, if we put $\xi_1 = f(1)$, $\xi_2 = f(2), \ldots$, we see that E_∞ is the space of all functions defined on the integers $1, 2, \ldots$.

EVENT SPACES

A very important kind of space that is considered in probability problems is an *event* space, that is, the space of all events that might occur in a given situation. For example, if two coins are tossed, the event space would contain the following events: the coins fall both heads, the coins fall both tails, the first coin falls heads and the second tails, and finally, the first coin falls tails and the second heads.

It is clear that the event space is neither a linear vector space nor a function space. Let us consider for a moment only event spaces which have a countable number of elements. Usually, to each event of such an event space is assigned a non-negative real number called its *weight*. For example, in the event space of the two tossed coins, we may assign the weight unity to each of the events. If the total sum of the weights is finite, we divide the weight assigned to each event by the total sum of the weights, thus obtaining the normalized weight or *probability* for that event. In the example of the two tossed coins, since the total sum of the weights is four, this method gives a probability of 1/4 for each event.

If the event space does not have a countable number of elements, then weights are assigned not to each event but to a family of subsets of the event space. For example, if any infinitely thin needle is spun about an axis through its center of gravity and perpendicular to the needle, then the events correspond to the final position of the needle; the final position is determined by the angle θ between 0 (inclusive) and 2π (exclusive) which the needle makes with a fixed line in the plane perpendicular to the axis. We assign a weight not to an individual event (which is specified by the value of θ) but to the subset of events corresponding to values of θ between two fixed numbers α and β. A possible weight for this subset of events would be $\beta - \alpha$; the total sum of weights is then found by integration to be 2π, and then the normalized weight or probability of this subset, that is, the probability of θ being between α and β, would be $(\beta - \alpha)/2\pi$.

To obtain a linear vector space we consider mappings of the event space into E_1, that is, assignments of real numbers to every event in the event space. This assignment defines a function on the event space. For example, in the case of the two tossed coins, assign to each event the number N of heads in the event. If HH, TT, HT, and TH denote the events, we may represent this assignment as follows:

$$N(\text{HH}) = 2, \qquad N(\text{TT}) = 0, \qquad N(\text{HT}) = N(\text{TH}) = 1.$$

Any real-valued function on an event space is called a *random variable*. Note that a random variable assumes its values with a certain normalized weight or probability. For the example, considered above, N is a random variable which assumes the values two and zero with probability 1/4 each, and the value one with probability $1/4 + 1/4 = 1/2$.

Since the space of random variables is a function space on the space of events, it is clear that the space of random variables is a linear vector space.

INNER PRODUCT

In vector analysis, that is, in the study of three-dimensional vectors, it is found that the concept of *inner* or *scalar* product of two vectors has three useful properties. The inner product of a vector with itself is the square of the length of the vector. The inner product of two vectors divided by the product of their lengths is the cosine of the angle between them. Two vectors are perpendicular (or equivalently, orthogonal) if and only if, their inner product is zero.

We shall try to introduce such a concept into an abstract vector space. Let S be a linear vector space containing elements $x, y, z, \ldots$. Consider a mapping of pairs of vectors into E_1, that is, an assignment of a real number to every pair of vectors x, y. We shall denote this real number by the symbol $\langle x, y \rangle$. This mapping defines an inner product if it has the following properties:

$$\langle x, y \rangle = \langle y, x \rangle, \tag{1}$$

$$\langle \alpha x_1 + \beta x_2, y \rangle = \alpha \langle x_1, y \rangle + \beta \langle x_2, y \rangle, \tag{2}$$

$$\langle x, x \rangle > 0, \quad \text{if } x \neq 0. \tag{3}$$

In E_3 the well-known definition of the inner product is as follows. If $x = (\xi_1, \xi_2, \xi_3)$ and $y = (\eta_1, \eta_2, \eta_3)$ then

$$\langle x, y \rangle = \xi_1 \eta_1 + \xi_2 \eta_2 + \xi_3 \eta_3. \tag{4}$$

It is clear that this definition satisfies the above properties (1), (2), and (3). Notice, however, that if α, β, γ are any real positive numbers then another possible definition for an inner product is

$$\langle x, y \rangle_1 = \alpha \xi_1 \eta_1 + \beta \xi_2 \eta_2 + \gamma \xi_3 \eta_3. \tag{5}$$

To understand the meaning of these different definitions, we introduce the concepts of the length of a vector and the orthogonality of two vectors. The *length* of a vector x is defined to be the positive square root of $\langle x, x \rangle$ and we write

$$|x| = \sqrt{\langle x, x \rangle}. \tag{6}$$

Two vectors x and y are said to be *orthogonal* or *perpendicular* if

$$\langle x, y \rangle = 0. \tag{7}$$

We shall call the vectors $e_1 = (1, 0, 0)$, $e_2 = (0, 1, 0)$, and $e_3 = (0, 0, 1)$ *base vectors* for E_3. The differences between the definitions (4) and (5) for the inner product can be stated as follows: in both (4) and (5) the base vectors are mutually orthogonal; in (4) they are of unit length but in (5) the length of e_1 is $\sqrt{\alpha}$, of e_2 is $\sqrt{\beta}$, and of e_3 is $\sqrt{\gamma}$.

We can imagine a more general situation in E_3 for which the base vectors are no longer mutually orthogonal. Then an appropriate inner product would be

$$\langle x, y \rangle_2 = \sum_{i=1}^{3} \sum_{j=1}^{3} \alpha_{ij} \xi_i \eta_j, \tag{8}$$

where the numbers α_{ij} are such that

$$\alpha_{ij} = \alpha_{ji} \tag{9}$$

and

$$\sum_{i=1}^{3} \sum_{j=1}^{3} \alpha_{ij}\, \xi_i\, \xi_j > 0 \tag{10}$$

if not all the ξ_i are zero. Condition (9) is a consequence of property (1) and condition (10) is a consequence of property (3).

It is clear that what has been said about inner products in E_3 holds also for inner products in E_n $(n = 1, 2, 3, \ldots)$. Suppose $x = (\xi_1, \xi_2, \ldots, \xi_n)$ and $y = (\eta_1, \eta_2, \ldots, \eta_n)$ are vectors in E_n, then we define

$$\langle x, y \rangle = \xi_1 \eta_1 + \ldots + \xi_n \eta_n. \tag{11}$$

We shall call E_n, with this definition of scalar product, an *n-dimensional Euclidean* space.

Similarly, if $x = (\xi_1, \xi_2, \ldots)$ and $y = (\eta_1, \eta_2, \ldots)$ are vectors in E_∞, we would like to define the inner product as follows:

$$\langle x, y \rangle = \xi_1 \eta_1 + \xi_2 \eta_2 + \ldots. \tag{12}$$

However, the difficulty is that the infinite series on the right side of (12) may not converge. We avoid this difficulty by considering only the set of vectors of finite length, that is, vectors x such that

$$\sqrt{\langle x, x \rangle} = \sqrt{\xi_1^2 + \xi_2^2 + \ldots} \tag{13}$$

is finite. If x and y are vectors of finite length then the series is (12) must converge absolutely. This follows from the obvious inequality

$$2\,|\xi_k \eta_k| \le \xi_k^2 + \eta_k^2$$

because then

$$2 \sum_{1}^{\infty} |\xi_k \eta_k| \le \sum_{1}^{\infty} \xi_k^2 + \sum_{1}^{\infty} \eta_k^2 < \infty. \tag{14}$$

From this it also follows that the sum of any two such vectors is also a vector of finite length, and that we are dealing with a linear vector space. Thus if x and y are vectors of finite length, then the length of $x + y$ is the square root of

$$\sum_{1}^{\infty} (\xi_k + \eta_k)^2 = \sum_{1}^{\infty} \xi_k^2 + \sum_{1}^{\infty} \eta_k^2 + 2 \sum_{1}^{\infty} \xi_k \eta_k$$

$$\leq 2 \sum \xi_k^2 + 2 \sum \eta_k^2 < \infty$$

by the use of (14). Henceforth, we shall use E_∞ to represent the linear vector space of vectors which have an infinite number of components and which have a finite length as defined in (13).

We give just one more example of an inner product. Suppose L_2 is the space of all real-valued functions $f(t)$, where $0 \leq t \leq 1$, which are square integrable, that is, such that

$$\int_0^1 f(t)^2 \, dt < \infty. \tag{15}$$

Then L_2 is a linear vector space with the inner product

$$<f, g> = \int_0^1 f(t) \, g(t) \, dt. \tag{16}$$

SCHWARZ INEQUALITY; PROJECTION

Let us return to the consideration of a linear vector space S in which there exists some inner product which has the properties (1), (2), and (3). We shall prove that any such inner product satisfies an important inequality called the *Schwarz inequality*.

We proceed in a geometric manner. We call $|x|$ the length of x, and it is defined as before by the positive square root of $<x, x>$. Suppose y is a nonzero vector; then we may ask what is the projection of x on the vector y. By the *projection*, we mean a vector αy such that the vector $x - \alpha y$ has the shortest possible length. [See Figure 2.1.]

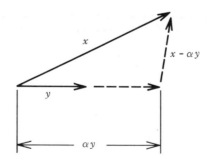

Figure 2.1

The squared length of $x - \alpha y$ is

$$<x - \alpha y, x - \alpha y> = <x, x> - 2\alpha <x, y> + \alpha^2 <y, y>. \tag{17}$$

If we differentiate this with respect to α and set the derivative equal to zero, we find that the length of $x - \alpha y$ will be a minimum for the value of α given by the equation

$$\alpha = <x,y>/<y,y> . \tag{18}$$

Substituting this value of α into (17), and using (6), we obtain the relation

$$|x - \alpha y|^2 = |x|^2 - |\alpha y|^2 , \tag{19}$$

or

$$|x|^2 = |x - \alpha y|^2 + |\alpha y|^2 . \tag{20}$$

The relation (20) has a very simple geometric meaning. Notice first that because of (18)

$$<x,y> - \alpha<y,y> = <x - \alpha y,y> = 0, \tag{21}$$

that is, the vector $x - \alpha y$ *is perpendicular to the vector* y. Thus (20) just expresses the Pythagorean theorem that the sum of the squares of the lengths of two mutually orthogonal vectors equals the square of the length of their vector sum.

Another interesting geometric fact can be obtained from (19). Since the length of a vector is non-negative, it follows that

$$|x|^2 \geq \alpha^2 |y|^2$$

or, by the use of (18), that

$$|x|^2 \cdot |y|^2 \geq <x, y>^2 .$$

This last result is the *Schwarz inequality*. It is more usually written as follows:

$$|<x, y>| \leq |x| \cdot |y| ; \tag{22}$$

that is, *the scalar product of two vectors is equal to or less than the product of the lengths of the vectors.*

This result should not surprise us, since in E_3 the scalar product of two vectors is equal to the product of the length of the vectors multiplied by the cosine of the angle between them and we know that the cosine is always less than one. By analogy, we may define the angle θ between two vectors x and y in an abstract space S by putting

$$\cos \theta = \frac{<x, y>}{|x| \cdot |y|} .$$

The most interesting case is that in which the angle is a right angle, and that occurs if and only if

$$<x,y> = 0 .$$

APPLICATIONS TO PROBABILITY

Let $E_1, E_2, \ldots$ be events in an event space with the corresponding non-negative weights $w_1, w_2, \ldots$. Then the total weight is

$$W = \sum_k w_k$$

and the probability of the event E_k is $P_k = w_k/W$. Consider the space of random variables, that is, the space of functions mapping events into numbers. We have seen that the space of random variables is a linear vector space. Suppose x and y are random variables such that x has the value $x(E_k)$ and y the value $y(E_k)$ when the event E_k occurs. We define the inner product of x and y as follows:

$$<x,y> = \sum_k P_k x(E_k) y(E_k) . \tag{23}$$

It is customary in probability theory to call $<x,y>$ the *expectation* of the product xy and to denote it by the symbol $E(xy)$; thus

$$E(xy) = <x,y> . \tag{24}$$

If the event space has only a finite number of elements, there is no difficulty with (23). However, if the event space contains an infinite number of elements, the infinite series in (23) may not converge. In that case we proceed as in E_∞ by restricting ourselves to random variables such that

$$<x,x> = E(x^2) \tag{25}$$

is finite. In probability theory $E(x^2)$ is called the *second moment* of the random variable and $E(xy)$ is called the *covariance* of the random variables x and y. From the Schwarz inequality (22), we immediately conclude that

$$E(xy)^2 \le E(x^2) E(y^2) ,$$

that is, the square of the covariance of x and y is less than the product of the second moment of x by the second moment of y.

A particularly important random variable is the one which assigns the number one to every event. We denote this random variable by the symbol **1**. Note that

$$E(1^2) = <1,1> = \sum_k P_k = 1$$

because of the way in which probability was introduced. We define

$$E(x) = E(1x) = <1,x> = \sum P_k x(E_k) . \tag{26}$$

This number $E(x)$ is called the *expected value* of x, the *first moment* of x, or the *mean value* of x. Because of the linearity of the inner product we have

$$E(x_1 + x_2 \ldots + x_r) = E(x_1) + E(x_2) + \ldots + E(x_n), \qquad (27)$$

that is, *the expected value of a sum of random variables equals the sum of the expected values*.

The projection of the random variable x on the random variable 1 is $E(x)1$ because $x - E(x)1$ is orthogonal to 1. This is shown as follows:

$$\langle x - E(x)1, 1 \rangle = \langle x, 1 \rangle - E(x)\langle 1, 1 \rangle = 0, \qquad (28)$$

by (26). We put $x' = x - E(x)1$ and we call x' the *normalized* variable x. Notice that from (28)

$$E(x') = 0. \qquad (29)$$

The second moment of x' is called the *variance* of the random variable x. It is denoted by $\sigma^2(x)$, where $\sigma(x)$ is called the *standard deviation* of x.

Since x' is orthogonal to 1, the Pythagorean theorem gives

$$\langle x', x' \rangle = \langle x, x \rangle - \langle x - x', x - x' \rangle.$$

By the use of the definition of x' and σ^2, this becomes

$$\sigma^2(x) = E(x'^2) = E(x^2) - E(x)^2. \qquad (30)$$

Equation (30) may be expressed in words thus: the variance of a random variable is the difference between the second moment and the square of the first moment. Notice that a zero variance implies $x' = 0$, which implies that x is a random variable that has the constant value $E(x)$ for all events.

Two random variables x and y are said to be *uncorrelated* if the corresponding normalized variables x' and y' are orthogonal, that is, if

$$E(x'y') = 0.$$

The *correlation coefficient* ρ between x and y is defined as

$$\rho = \frac{E(x'y')}{[E(x'^2)E(y'^2)]^{\frac{1}{2}}}. \qquad (31)$$

By the Schwarz inequality it follows that

$$-1 \leq \rho \leq 1. \qquad (32)$$

Note that $\rho = 0$ implies that x and y are uncorrelated.

Suppose that the random variables $x_1, x_2, \ldots, x_n$ are mutually uncorrelated, that is, suppose for $j \neq k$,

$$E(x'_j x'_k) = 0, (j, k = 1, 2, \ldots, n).$$ (33)

Then the variance of the sum $x_1 + \ldots + x_n$ is the sum of the variances; to prove this, notice that

$$\sigma^2 (x_1 + \ldots + x_n) = E[(x'_1 + \ldots + x'_n)^2]$$

$$= \sum_{j=1}^{n} E\,(x'^2_j) + \sum_{j \neq k}\sum E(x'_j x'_k)$$

$$= \sum_1^n E(x'^2_j) = \sum_1^n \sigma^2(x_j)$$ (34)

because of (33).

We shall illustrate these concepts by considering the tossing of n coins. The possible events are the different arrangements of heads and tails that may occur. Clearly, since each coin may fall heads (H) or tails (T) independent of every other, the total number of events is 2^n. We shall assign a weight one to each event; then the probability of a preassigned arrangement of H and T is just 2^{-n}.

Let N denote the number of H's in an event. What is the expected value of N? For example, if $n = 2$ the events are HH, HT, TH, and TT and the random variable N is defined as follows:

$$N(\text{HH}) = 2, \quad N(\text{HT}) = N(\text{TH}) = 1, \quad N(\text{TT}) = 0.$$

By the definition (26) we find that

$$E(N) = \frac{1}{4} \cdot 2 + \frac{1}{4} \cdot 1 + \frac{1}{4} \cdot 1 + \frac{1}{4} \cdot 0 = 1.$$

It would be quite difficult to do the general case in this way. Instead, we introduce random variables $x_1, x_2, \ldots, x_n$ which are defined as follows: $x_j = 1$ if the jth coin is H and $x_j = 0$ if the jth coin is T. We see that

$$N = x_1 + \ldots + x_n.$$

We have shown that

$$E(N) = E(x_1) + \ldots + E(x_n).$$

Since the jth coin is H with probability $\frac{1}{2}$, we find that $E(x_j) = \frac{1}{2}$ and therefore $E(N) = n/2$.

We may also find the variance of N in a similar way. The random variables $x_1, \ldots, x_n$ are easily shown to be mutually uncorrelated; therefore, by (34)

$$\sigma^2(N) = \sigma^2(x_1) + \ldots + \sigma^2(x_n).$$

Since by (30)

$$\sigma^2(x_j) = E(x_j{}^2) - E(x_j)^2 = 1 \cdot \frac{1}{2} - \left(\frac{1}{2}\right)^2 = \frac{1}{4},$$

we conclude that

$$\sigma^2(N) = \frac{n}{4}.$$

2.2 LINEAR OPERATORS

Let S be a linear vector space with an inner product. An *operator L* is a mapping of the elements x of S into the elements of Lx of another linear vector space S'. We shall usually restrict ourselves to the case where S' is identical with S. For example, if S is the space of continuous functions $f(t)$, $g(t)$ where $0 \le t \le 1$, then the mapping of $f(t)$ into the function $Lf(t) = g(t)$ is an operator on S into S.

An operator L is called *linear* if

$$L(\alpha x) = \alpha\, Lx,$$

where α is any constant, and

$$L(x + y) = Lx + Ly.$$

The following illustrations of linear operators include matrices, integral operators, differential operators, and difference operators.

(a) In E_n with elements $x = (\xi_1, \xi_2, \ldots, \xi_n)$, let L be a matrix with elements a_{ij} $(i, j = 1, 2, \ldots, n)$; then the operator Lx which maps x into a vector x' in E_n with components $\xi_i (i = 1, 2, \ldots, n)$, where

$$\xi_j' = \sum_{j=1}^{n} a_{ij}\, \xi_j,$$

is a linear operator.

(b) Let S be the space of all continuous functions $f(t)$ for $0 \le t \le 1$; then the operator defined by the integral

$$Lf(t) = \int_0^1 k(t,s) f(s)\, ds,$$

where $k(t,s)$ is a continuous function of both s and t, is a linear operator.

(c) Let S be the space of all functions $f(t)$ defined for $0 \le t \le \infty$ which are continuous and have continuous first and second derivatives. Then the operator defined by the formula

$$Lf(t) = f''(t + 1) + 7f'(2t) + f(t)$$

is a linear operator.

An operator L is *continuous* if, whenever a sequence of vectors x_n converges to a vector x, then the vectors Lx_n converge to the vector Lx. The operator is bounded if there exists a constant γ such that

$$|Lx| < \gamma |x|$$

for all x in S. It can be proved that a bounded operator is continuous and that, conversely, a continuous operator is bounded; this makes it easy to derive various useful properties of the class of bounded operators. Unfortunately, however, all differential operators are unbounded, and as a consequence it is much harder to prove the desired results for such operators.

To illustrate the fact that differential operators are unbounded, let us consider the space of functions $f(t)$, defined for $0 \leq t \leq 1$, which are continuous and continuously differentiable. Put $Lf(t) = f'(t)$ and consider the action of L on the function $\sin n\pi t$. We have

$$\langle Lf, Lf \rangle = |L(\sin n\pi t)|^2 = \int_0^1 n^2\pi^2 \cos^2 n\pi t \, dt = n^2\pi^2/2,$$

but

$$\langle f, f \rangle = \langle \sin n\pi t, \sin n\pi t \rangle = \int_0^1 \sin^2 n\pi t \, dt = \frac{1}{2};$$

consequently, the ratio

$$|Lf(t)| / |f(t)| = n\pi,$$

and it is clear that this ratio cannot be bounded.

ADJOINT OPERATOR

It is convenient to associate with an operator L another operator L^* called its *adjoint*. The adjoint operator is defined as follows.

For any two vectors x, y in the space, consider the scalar product $\langle y, Lx \rangle$. We try to write it as the scalar product of x with another vector w, that is,

$$\langle y, Lx \rangle = \langle w, x \rangle, \tag{35}$$

such that

$$w = L^*y .$$

The fundamental property of the *adjoint operator* is thus that

$$\langle y, Lx \rangle = \langle L^*y, x \rangle . \tag{36}$$

The adjoint operator is obtained in particular cases by using the formula (35). For example, in case (a) considered above, if we put $y = (\eta_1, \ldots, \eta_n)$, we have

$$<y, Lx> = \sum_i \sum_j \eta_i a_{ij} \xi_j = \sum_j \xi_j \sum_i a_{ij} \eta_i ,$$

and by comparing with (36), we see that

$$L^*y = \sum_i a_{ij} \eta_i ;$$

that is, the operator L^* is represented by the matrix which is the transpose of the matrix that represents L.

In case (b) above, let $g(t)$ be another function in the space; then

$$<g, Lf> = \int_0^1 g(t)\, dt \int_0^1 k(t, s) f(s)\, ds$$

$$= \int_0^1 f(s)\, ds \int_0^1 k(t, s) g(t)\, dt .$$

Again comparing with (36), we see that

$$L^*g = \int_0^1 k(s, t) g(s)\, ds .$$

In the case of differential operators, the adjoint operator is obtained by the method of integration by parts. Suppose that S is the space of functions $f(t)$ defined on the interval $0 \leq t \leq 1$ such that $f(t)$ and its first and second derivatives are continuous, and suppose also that $f(0) = f(1) = 0$. Let $Lf(t) = f''(t)$. To find the adjoint operator L^*, let $g(t)$ be a similar function in S; then

$$<g, Lf> = \int_0^1 g(t) f''(t)\, dt$$

$$= [g(t) f'(t) - f(t) g'(t)]_0^1 + \int_0^1 g''(t) f(t)\, dt$$

$$= \int_0^1 g''(t) f(t)\, dt ,$$

because, by assumption,

$$f(0) = f(1) = g(0) = g(1) = 0 .$$

We see that

$$L^*g = g''(t) .$$

An operator such as this one, for which $L = L^*$, is called *self-adjoint*. Note that because of (36), if L is self-adjoint, then

$$< y , Lx > \ = \ < Ly , x > . \tag{37}$$

It is easy to see that the adjoint of a linear combination of operators is the same linear combination of the adjoints, but we shall show that the adjoint of a product of two operators L and M is the product of the adjoints in the reverse order, that is, that

$$(LM)^* \ = \ M^* L^*. \tag{38}$$

This follows from

$$< (LM)^* y , x > \ = \ < y , LMx > \ = \ < y , L(Mx) > \ = \ < L^* y , Mx >$$
$$= \ < M^* L^* y , x > .$$

QUANTUM MECHANICS

In naive quantum mechanics the *state* of a dynamical system, such as an electron bound to an atom, is represented by an element of a Hilbert space. The elements of the Hilbert space are complex quantities and the appropriate scalar product is of the complex type. This means that now the scalar product is no longer symmetric but instead, if ψ and ϕ are elements of the Hilbert space, then

$$< \psi , \phi > \ = \ \overline{< \phi , \psi >} , \tag{39}$$

where the bar denotes the complex conjugate.

A *dynamical variable*, such as the energy or momentum, is represented by a linear operator on the Hilbert space. If the operator is self-adjoint, then the corresponding dynamical variable is called an *observable*. The *average value* of an observable L in the state ϕ is defined to be

$$\frac{< \phi , L \phi >}{< \phi , \phi >} . \tag{40}$$

Because L is a self-adjoint operator, we can show that the average value of L must be a real number. From (39), we have that

$$< L\phi , \phi > \ = \ \overline{< \phi , L\phi >} \ = \ \overline{< L\phi , \phi >}$$

by (37); consequently, the scalar product $< \phi , L\phi > \ = \ \overline{< \phi , L\phi >}$ is a real number. Since $< \phi , \phi >$ is a nonnegative real number, this proves that the average value of L is a real number.

Formula (40) takes a simpler form if we assume that ϕ is normalized so that $< \phi , \phi > \ = \ 1$. In that case the average value of L is $< \phi , L\phi >$. We may also define the average value of a function of an operator $f(L)$ as $< \phi , f(L)\phi >$. For example, if the average of L in the state ϕ is λ, then the average value of $L - \lambda$ in the state ϕ is

$$<\phi, (L-\lambda)\phi> \ = \ <\phi, L\phi> \ - \ \lambda<\phi, \phi> \ = \ 0 \ ;$$

but the average value of $(L-\lambda)^2$ in the state ϕ is

$$<\phi, (L-\lambda)^2\phi> \ = \ <(L-\lambda)\phi, (L-\lambda)\phi> \ = \ |(L-\lambda)\phi|^2 \ , \tag{41}$$

a positive quantity unless $(L-\lambda)\phi = 0$.

A state ϕ such that $(L-\lambda)\phi = 0$ is called a *pure state* or an *eigenstate* for the operator ϕ. For an eigenstate ϕ, we have $L^2\phi = L(L\phi) = \lambda L\phi = \lambda^2\phi$, and in general

$$f(L)\phi \ = \ f(\lambda)\phi \ .$$

From this it follows that the average value of any function $f(L)$ in the state ϕ is $f(\lambda)$. This means that in the state ϕ the observable L always has the value λ (the eigenvalue).

If ϕ is not an eigenstate of L, then the quantity

$$\sigma^2 \ = \ |(L-\lambda)\phi|^2 \tag{42}$$

is positive and different from zero. We have

$$\sigma^2 \ = \ <(L-\lambda)\phi, (L-\lambda)\phi>$$

$$= \ <L\phi, L\phi> \ - \ <\lambda\phi, L\phi> \ - \ <L\phi, \lambda\phi> \ + \ |\lambda|^2<\phi, \phi>$$

$$= \ |L\phi|^2 \ - \ \lambda^2 \ ,$$

by the definition of λ. This shows that the average value of L^2 is not the same as the square of the average of L:

$$<\phi, L^2\phi> \ = \ <L\phi, L\phi> \ = \ \lambda^2 + \sigma^2 ,$$

and consequently, the observable L does not always have the value λ when measured on the state. The quantity σ^2 is called the *fluctuation* or the *variance* of the observable L.

From (40) and (42) we conclude that when L is measured on a state ϕ, the results form a random variable whose average is λ and whose variance is σ^2.

UNCERTAINTY PRINCIPLE

The previous section has shown that an observable L cannot be measured exactly in a state ϕ unless ϕ is an eigenstate of L. Suppose we have two observables A and B. When can they be simultaneously measured exactly? We shall show that if the observables A and B do not commute, then there exists a fundamental limitation to the accuracy of their simultaneous measurement.

We shall suppose that A and B satisfy the following relation:

$$AB - BA = \gamma I, \tag{43}$$

where γ is a scalar and I is the identity operator; that is, $I\phi = \phi$ for all ϕ. If we take the adjoint of equation (43) and use the fact that A and B are self-adjoint, we find that

$$BA - AB = \overline{\gamma I} = -\gamma I \, ;$$

this implies that $\overline{\gamma} = -\gamma$; consequently, γ is a pure imaginary number.

Let ϕ be a normalized state and suppose the average value of A and B on ϕ are α and β, respectively. Put

$$A' = A - \alpha I, \quad B' = B - \beta I,$$

so that the average value of both A' and B' on ϕ is zero. If in (43) we put $A = A' + \alpha I$ and $B = B' + \beta I$, we obtain the similar relation

$$A'B' - B'A' = \gamma I. \tag{44}$$

With the help of the Schwarz inequality, we find that

$$|<\phi, A'B'\phi>| = |<A'\phi, B'\phi>| \le |A'\phi| \cdot |B'\phi|$$

Using this inequality and (44), we obtain

$$|<\phi, \gamma \phi>| = |<\phi, A'B'\phi> - <\phi, B'A'\phi>|$$

$$\le |A'\phi| \cdot |B'\phi| + |B'\phi| \cdot |A'\phi|$$

$$= 2|A'\phi| \cdot |B'\phi|,$$

or

$$|A'\phi|^2 \cdot |B'\phi|^2 \ge \frac{1}{4}\gamma^2. \tag{45}$$

Since

$$|A'\phi|^2 = |(A - \alpha I)\phi|^2$$

is the variance of the observable A, and corresponds to the error inherent in measuring A in the state ϕ, formula (45) shows that the product of the error in measuring A by the error in measuring B cannot be less than the constant value $\gamma/2$. This means that if the state ϕ is so chosen that A can be measured with great accuracy (i.e., small variance), then the error in measuring B in this state is correspondingly large. We conclude that if A and B do not commute, they cannot be simultaneously measured accurately.

SPECTRAL REPRESENTATION

The previous section has shown the importance of the eigenstates of an observable in quantum mechanics. In this section we consider the extension of this concept to arbitrary linear operators in a linear vector space S.

Let L be a linear operator on S. A nonzero vector x is called an *eigenvector* of L if

$$Lx = \lambda x, \tag{46}$$

where λ is a scalar. This scalar λ is called an *eigenvalue* of the operator L. Note that the action of L on the eigenvector x is just simple multiplication by the scalar λ. This is illustrated by

$$L^k x = L^{k-1}(Lx) = \lambda L^{k-1} x = \ldots = \lambda^k x.$$

Suppose the operator L is self-adjoint and let x and y be eigenvectors of L corresponding to the eigenvalues of λ and μ, respectively. We shall prove that if $\lambda \neq \mu$, then the eigenvectors x and y are orthogonal. From the definition of eigenvector we have

$$Lx = \lambda x, \qquad Ly = \mu y;$$

then, using the fact that L is self-adjoint, we get

$$\lambda \langle y, x \rangle = \langle y, Lx \rangle = \langle Ly, x \rangle = \mu \langle y, x \rangle,$$

or

$$(\lambda - \mu) \langle y, x \rangle = 0.$$

Since $\lambda - \mu \neq 0$, this implies

$$\langle y, x \rangle = 0.$$

For some operators, the set of its eigenvectors $x_1, x_2, \ldots$, corresponding to eigenvalues $\lambda_1, \lambda_2, \ldots$, may form a basis for (and thereby span) the space. This means that, given any x in the space, there exist scalars $\alpha_1, \alpha_2, \ldots$ such that

$$x = \Sigma \alpha_k x_k \tag{47}$$

and

$$Lx = \Sigma \lambda_k \alpha_k x_k. \tag{48}$$

We call formulas (47) and (48) the *spectral representation* of the operator L.

If L is self-adjoint, there is a very simple method for finding the coefficients α_k in (47). We have proved previously that eigenvectors corresponding to different eigenvalues are orthogonal. Consider a set of eigenvectors, say, $x_1, x_2, \ldots, x_n$, which all correspond to the *same* eigenvalue λ. This means that

$$Lx_1 = \lambda x_1, \ldots, Lx_n = \lambda x_n;$$

but then

$$L(a_1 x_1 + \ldots + a_n x_n) = \lambda(a_1 x_1 + \ldots + a_n x_n).$$

This shows that every vector in the linear subspace spanned by x_1, x_2, $\ldots$, x_n is an eigenvector of L corresponding to the same eigenvalue λ. In this subspace we may introduce an orthonormal basis by the following construction.

Pick any vector of unit length y_1, in this subspace. Suppose y_2 is a vector in this subspace which is linearly independent of y_1 (geometrically, noncollinear with y_1); then consider y_2 minus its projection on y_1, that is, the vector

$$y_2 - \langle y_2, y_1 \rangle y_1.$$

This vector is orthogonal to y_1 because

$$\langle y_1, (y_2 - \langle y_2, y_1 \rangle y_1) \rangle = \langle y_1, y_2 \rangle - \langle y_2, y_1 \rangle \langle y_1, y_1 \rangle = 0.$$

Put

$$z_2 = \beta[y_2 - \langle y_2, y_1 \rangle y_1],$$

where β is so chosen that z_2 has the length one. The vectors y_1, z_2 are the first two vectors in the desired orthonormal basis. We continue by choosing a vector y_3 that is linearly independent of both y_1 and z_2 (geometrically, noncoplanar with y_2 and z_2); then consider the vector that is y_3 minus its projection on the subspace spanned by y_1 and z_2, that is, the vector

$$y_3 - \langle y_3, y_1 \rangle y_1 - \langle y_3, z_2 \rangle z_2.$$

Clearly, this vector is orthogonal to both y_1 and z_2. Put

$$z_3 = \gamma[y_3 - \langle y_3, y_1 \rangle y_1 - \langle y_3, z_2 \rangle z_2],$$

where γ is so chosen such that z_3 has length one. The vectors y_1, z_2, and z_3 are the first three vectors of the orthonormal basis. It is clear that this process can be continued until a complete orthonormal basis of eigenvectors of L is obtained.

Let us recapitulate. Eigenvectors corresponding to different eigenvalues are mutually orthogonal. Eigenvectors corresponding to the same eigenvalue can be replaced by an orthonormal set of eigenvectors; consequently, the set of eigenvectors of the self-adjoint operator L form an orthonormal basis for the space S. Suppose that x_1, x_2, $\ldots$ are the vectors of this orthonormal basis. This implies that

$$\langle x_i, x_j \rangle = \delta_{ij}, \tag{49}$$

where $\delta_{ij} = 1$, if $i = j$, but $\delta_{ij} = 0$ otherwise. With the help of (49), it is easy to find the coefficients a_k in (47). In fact, we have

$$\langle x_j, x \rangle \ = \ \Sigma \, \alpha_k \langle x_j, x_k \rangle \ = \ \Sigma \, \alpha_k \delta_{jk} \ ;$$

therefore,

$$\alpha_j \ = \ \langle x_j, x \rangle . \qquad (50)$$

A simple illustration may help clarify the meaning of formulas (47), (48), and (50). Let L be the operator $-d^2/dt^2$ on the domain of functions $u(x)$, which are continuous, which have continuous first derivatives and piecewise continuous second derivatives on the interval $0 \le t \le 1$, and which are such that $u(0) = u(1) = 0$. An eigenvector, or eigenfunction, of L is a nonzero function $u(t)$ in the domain of L such that

$$Lu(t) \ = \ -u''(t) \ = \ \lambda u(t). \qquad (51)$$

The solutions of this equation are $\sin \sqrt{\lambda} t$ and $\cos \sqrt{\lambda} t$. Because of the boundary condition at $t = 0$, we must take

$$u(t) \ = \ \alpha \sin \sqrt{\lambda} t \ ,$$

where α is an arbitrary scalar; however, the boundary condition at $t = 1$ implies that

$$\alpha \sin \sqrt{\lambda} \ = \ 0 .$$

Since $\alpha = 0$ would make $u(t)$ identically zero and is therefore inadmissible, we require $\sin \sqrt{\lambda} = 0$. The roots of this equation are

$$\lambda \ = \ (n\pi)^2 , \qquad n \ = \ 1, \ 2, \ 3, \ \ldots .$$

For each value of n, there exist an eigenfunction $\sin n\pi t$ and the corresponding eigenvalue $n^2\pi^2$. If we put

$$u_n(t) \ = \ \sqrt{2} \ \sin n \pi t \ , \qquad (52)$$

then the eigenfunctions $u_n(t)$ from an orthonormal set because

$$\int_0^1 u_n(t) \, u_m(t) \, dt \ = \ 2 \int_0^1 \sin n\pi t \ \sin m\pi t \ dt \ = \ \delta_{nm} . \qquad (53)$$

From the theory of Fourier series it is known that these functions (52) form a basis for the space; that is, given any function $u(t)$ which is square-integrable over the interval $(0, 1)$, there exist scalars $\alpha_1, \alpha_2, \ldots$ such that

$$u(t) \ = \ \sum \alpha_k \sin k\pi t \ .$$

In fact, by the use of (53) it is easy to obtain the following well-known formula for the Fourier coefficients:

$$\alpha_k \ = \ 2 \int_0^1 u(t) \sin k\pi t \ dt \ .$$

FUNCTIONS OF AN OPERATOR

The problem of solving an operator equation can be regarded as the problem of inverting and interpreting some function of the operator. For example, consider the problem of solving the equation

$$(L - \lambda)x = b ,$$

where b is a given vector and x is desired. The formal solution of this is

$$x = (L - \lambda)^{-1} b , \qquad (54)$$

but the difficulty of interpreting the meaning of the operator $(L - \lambda)^{-1}$ still remains.

The spectral representation of L, as defined in (47) and (48), can be used to interpret functions of L. Note that

$$L^2 x = L(Lx) = \sum \lambda_k^2 \, \alpha_k \, x_k ,$$

and more generally,

$$L^n x = \sum \lambda_k^n \, \alpha_k \, x_k .$$

If $p(L)$ is a polynomial in L, that is, if

$$p(L) = \alpha_0 + \alpha_1 L + \ldots + \alpha_n L^n ,$$

then we get

$$p(L)x = \sum p(\lambda_k) \, \alpha_k \, x_k .$$

A similar result is easily obtained for a function $\phi(\lambda)$ which is analytic for $\lambda = \lambda_1, \lambda_2, \ldots .$ We find that

$$\phi(L)x = \sum \phi(\lambda_k) \, \alpha_k \, x_k . \qquad (55)$$

Formula (55) may be applied to give a solution of (54). Since b is a known vector, we can find

$$\beta_k = <x_k, b> ,$$

and then, as for (47) and (50), we have

$$b = \sum \beta_k \, x_k .$$

Let us use (55) with $\phi(L) = (L - \lambda)^{-1}$ and $x = b$; then

$$(L - \lambda)^{-1} b = \sum \frac{\beta_k \, x_k}{\lambda_k - \lambda}$$

is the solution of (54).

For later use we shall obtain one more result. Consider a contour C in the complex λ plane such that C encloses the spectrum (the set λ_k) of L. Then by (55),

$$\frac{1}{2\pi i}\int_C d\lambda\,\frac{1}{L-\lambda}\,x = \sum\frac{1}{2\pi i}\int_C\frac{d\lambda}{\lambda_k-\lambda}\,\alpha_k x_k = -\sum\alpha_k x_k = -x,$$

since the residue at the poles $\lambda = \lambda_k$ is -1. We conclude that

$$\frac{1}{2\pi i}\int_C d\lambda\,\frac{1}{\lambda-L}\,x = x. \tag{56}$$

This formula was proved by the use of the spectral representation of L. However, we shall use it later to obtain the spectral representation.

CONTINUOUS SPECTRUM

The methods we have just considered for interpreting functions of an operator L depend on the fact that the eigenvectors of L for a basis for the space S. There exist, however, operators such that their eigenvectors do not form a basis for the space; consequently, the spectral representation given in (47) and (48) is not valid. Nevertheless, we shall show how the idea of the spectral representation can be extended to such operators.

We start by considering the operator $L = -d^2/dt^2$ on the domain of functions $u(t)$, defined for $0 \leqslant t < \infty$, which are continuous and have continuous first derivatives and piecewise continuous second derivatives, which are square-integrable over $(0,\infty)$, and which satisfy the boundary condition

$$u(0) = 0. \tag{57}$$

The eigenfunctions of L, if any, would be nonzero functions $u(t)$ in the domain of L such that

$$-u''(t) = \lambda u(t),$$

and would satisfy (57). Such functions are

$$u(t) = \alpha\,\sin\sqrt{\lambda}\,t,$$

where α is an arbitrary scalar. However, this function is not of integrable square because, no matter what the value of λ,

$$\int_0^\infty u(t)^2\,dt = \alpha^2\int_0^\infty\sin^2\sqrt{\lambda}\,t\,dt = \infty.$$

Therefore, the operator L has no eigenfunction and so (47) and (48) cannot be used.

To remedy the lack of eigenfunctions, we introduce the concepts of *continuum eigenvalue* and of *continuum eigenfunction*. A number λ is said to be a continuum eigenvalue for an operator L if there exists a sequence of functions $u_n(t)$ in the domain of L such that the ratio

$$\frac{|(L - \lambda)u_n|}{|u_n|} \tag{58}$$

converges to zero as n converges to infinity. If the functions $u_n(t)$ converge pointwise to a function $u(t)$, then $u(t)$ is called a continuum eigenfunction of L. (Note that if the functions $u_n(t)$ were to converge "in the sense of the space" to $u(t)$, then $u(t)$ would be an eigenfunction of L corresponding to the eigenvalue λ. In general, we shall restrict the term continuum eigenfunction to the case where there is no convergence in the sense of the space so that $u(t)$ is *not* an eigenfunction and λ is *not* an eigenvalue.) We shall emphasize the distinction between eigenvalues and continuum eigenvalues by saying that if λ is an eigenvalue, it belongs to the *discrete spectrum* of L, but if λ is a continuum eigenvalue, it belongs to the *continuous spectrum* of L.

For the particular operator $L = -d^2/dt^2$ considered above, we can show that every positive value of λ belongs to the continuous spectrum. To see this, consider the sequence of functions $u_n(t)$ defined as follows:

$$
\begin{aligned}
u_n(t) &= \sin \sqrt{\lambda}\, t, & 0 \le t \le \left(2n + \frac{1}{2}\right)\pi \lambda^{-1/2} = t_n \\[2mm]
&= 1 - 2(t - t_n)^2, & t_n \le t \le t_n + \frac{1}{2}, \\[2mm]
&= 2(t - t_n - 1)^2, & t_n + \frac{1}{2} \le t \le t_n + 1, \\[2mm]
&= 0 & t_n + 1 \le t
\end{aligned}
$$

Note that $u_n(t)$ is continuous, has continuous first derivatives and piecewise continuous second derivatives, is of integrable square, and satisfies the condition $u_n(0) = 0$; consequently, $u_n(t)$ is in the domain of L. We find that

$$
\begin{aligned}
(L - \lambda)\, u_n(t) &= 0, & 0 \le t \le t_n, \\[2mm]
&= 4 - \lambda[1 - 2(t - t_n)^2], & t_n \le t \le t_n + \frac{1}{2}, \\[2mm]
&= -4 - 2\lambda(t - t_n - 1)^2, & t_n + \frac{1}{2} \le t \le t_n + 1, \\[2mm]
&= 0 & t_n \le t.
\end{aligned}
$$

Evaluating the ratio (58), we see that it is less than

$$\frac{\displaystyle\int_0^{1/2} [4 - \lambda(1 - 2t)^2]^2\, dt + \int_{1/2}^1 [4 + 2\lambda(t - 1)^2]^2\, dt}{\displaystyle\int_0^{t_n} \sin^2 \sqrt{\lambda}\, t\ dt}.$$

Since the numerator is independent of n and since the denominator goes to infinity with n, we conclude that the ratio goes to zero and therefore λ is in the continuous spectrum. Obviously, the limit of $u(t)$ as n goes to infinity is the continuum eigenfunction $\sin \sqrt{\lambda}\, t$.

The continuum eigenfunctions $\sin \sqrt{\lambda}\, t$ for $\lambda > 0$ can be normalized in many ways. [For details see Friedman, Chp. 4.] The consequence of the normalization will be a generalized spectral representation for this operator L, as follows.

If $u(t)$ is a function in the domain of L, there exists a square-integrable function $g(k)$ such that

$$u(t) = \left(\frac{2}{\pi}\right)^{1/2} \int_0^\infty g(k) \sin kt \, dk \tag{59}$$

and

$$Lu(t) = \left(\frac{2}{\pi}\right)^{1/2} \int_0^\infty k^2 g(k) \sin kt \, dk . \tag{60}$$

Note that we have put $\lambda = k^2$. These formulas (59) and (60) are the analog of (47) and (48). There is also an analog of formula (50). It is

$$g(k) = \left(\frac{2}{\pi}\right)^{1/2} \int_0^\infty u(t) \sin kt \, dt . \tag{61}$$

We call (59) and (60) the spectral representation of the operator L.

It can be shown that for any self-adjoint operator L there exists a spectral representation. This spectral representation will in general contain both a discrete and a continuous spectrum; therefore, the function $u(t)$ will be represented as a combination of a sum, such as in (47), and an integral, such as in (59), but the action of L on $u(t)$ will be essentially a multiplication by the corresponding eigenvalue, discrete or continuum.

PARTIAL DIFFERENTIAL EQUATIONS

Separable partial differential equations can be solved by the use of the concepts of the function of a differential operator and its spectral representation. The idea is this: all the partial differential operators but one are considered to be scalars and the resulting ordinary differential equation is solved. The resulting solution is a function of the operators which were considered scalars. This function of the operators is interpreted by the use of the spectral representation of the operators, thus giving the solution of the original problem.

We shall illustrate this method by two examples. First, consider the problem of finding the potential produced by a line charge inside a hollow rectangular conductor. Let $W(x,y)$ be the required potential. It satisfies the equation

$$W_{xx} + W_{yy} = -\delta(x - x_0)\, \delta(y - y_0), \tag{62}$$

where (x_0, y_0) is the location of the line charge. Suppose that the dimensions of the conductor are a by b; then

$$W(0, y) = W(a, y) = W(x, 0) = W(x, b) = 0 . \tag{63}$$

To solve (62), we put

$$L = \frac{-d^2}{dx^2}$$

with the conditions that the operand, say $u(x)$, vanishes at $x = 0, a$; then we may write (62) as

$$\frac{d^2 W}{dy^2} - LW = -\delta(x - x_0)\, \delta(y - y_0) \tag{64}$$

with the boundary conditions that $W = 0$ for $y = 0, b$. The solution of (64) satisfying these boundary conditions is

$$W = \frac{\sinh \sqrt{L}\, y \, \sinh \sqrt{L}\, (b - y_0)}{\sqrt{L}\, \sinh \sqrt{L}}\, \delta(x - x_0), \quad y < y_0,$$

$$= \frac{\sinh \sqrt{L}\, (b - y)\, \sinh \sqrt{L}\, y_0}{\sqrt{L}\, \sinh \sqrt{L}}\, \delta(x - x_0), \quad y > y_0 . \tag{65}$$

To interpret this function of L, we use the spectral representation of δ; the eigenfunctions are similar to those given in (52). We have

$$\delta(x - x_0) = \frac{2}{a} \sum_1^\infty \sin\frac{k\pi x}{a} \sin\frac{k\pi x_0}{a} \tag{66}$$

and the eigenvalues of L are $(k\pi/a)^2$ for $k = 1, 2, \ldots$. Using (66) in (65), we find that

$$W = \frac{2}{\pi} \sum_1^\infty \frac{\sinh (k\pi y/a)\, \sinh k\pi(b - y_0)/a}{k \, \sinh (k\pi/a)} \sin\frac{k\pi x}{a} \sin\frac{k\pi x_0}{a}, \quad y < y_0, \tag{67}$$

and a similar result for $y > y_0$. Formula (67) is the solution of (62).

As a second illustration, we seek a function $W(x, y)$ such that

$$W_{xx} + W_{yy} = -\delta(x - x_0)\, \delta(y - y_0) \tag{68}$$

for $0 \le x < \infty$, $0 \le y \le b$, and such that $W(x, y)$ satisfies the boundary conditions

$$W(x, 0) = W(x, b) = W(0, y) = 0 .$$

Again, we put $L = -d^2/dx^2$, but now with the boundary conditions $u(0) = 0$ and $u(x)$ square-integrable over $(0, \infty)$. Equation (68) becomes

$$\frac{d^2W}{dy^2} - LW = - \delta(x - x_0)\, \delta(y - y_0),$$

with the condition that $W = 0$ for $y = 0$ and $y = b$. The solution of this equation is (65), and again the meaning of the function of the operator can be obtained from the spectral representation that is given in (59) and (60). Using it, we find that

$$\delta(x - x_0) - \frac{2}{\pi} \int_0^\infty \sin kx \sin kx_0 \, dk$$

and putting this in (65) we get the following solution for (62):

$$W = \frac{2}{\pi} \int_0^\infty \frac{\sinh ky \, \sinh k(b - y_0)}{k \sinh k} \, \sin kx \, \sin kx_0 \, dk, \quad y < y_0,$$

$$= \frac{2}{\pi} \int_0^\infty \frac{\sinh k(b - y) \, \sinh ky_0}{k \sinh k} \, \sin kx \, \sin kx_0 \, dk, \quad y > y_0.$$

Asymptotic Methods

INTRODUCTION

In the applications of mathematics to physical situations, it is not sufficient to obtain a mathematically satisfactory solution to a given problem; it is also necessary to devise a convenient method for obtaining numbers from the solution, and to construct relatively simple analytic expressions in which the dependence of the phenomena on the parameters of the problem is clear. Of course, because of the development of high-speed computers in recent years, the urgency for devising such methods for numerical purposes has been lessened. However, there still exist situations in which a comparatively inaccurate analytic expression is more useful than a very accurate table of numerical values, and there also exist situations in which the cost and effort involved in putting the problem on a high-speed computer are not justified by the results. Consequently, it is of practical as well as theoretical interest to study methods for approximating mathematical expressions.

We shall discuss certain methods for approximating the value of a function that is defined by an integral. It will turn out that the accuracy of those approximations is limited and cannot be improved beyond a definite amount. However, as the argument of the function increases to infinity, the error of the approximation goes to zero.

3.1 ASYMPTOTIC SERIES

ILLUSTRATION

To illustrate the problems involved, consider the function defined by the integral

$$F(x) = \int_0^\infty \frac{e^{-xt}}{1 + t} \, dt . \tag{1}$$

We shall show how $F(x)$ may be approximated, with a known estimate for the error, by certain simple expressions. These expressions may be used either to compute the function $F(x)$, or to replace $F(x)$ by a simpler function so that other analytical operations, such as integration, may be more easily carried out.

We begin by expanding the function $1/(1 + t)$ in a power series:

$$\frac{1}{1 + t} = \sum_{n=0}^\infty (-1)^n t^n . \tag{2}$$

Then, formally,

$$F(x) = \int_0^\infty \sum (-1)^n t^n e^{-xt} \, dt,$$ (3)

and since

$$\int_0^\infty t^n e^{-xt} \, dt = \frac{n!}{x^{n+1}},$$ (4)

we write

$$F(x) \sim \sum_{n=0}^\infty \frac{(-1)^n n!}{x^{n+1}} = \sum_{n=0}^\infty (-1)^n u_n,$$ (5)

where we use the symbol $\sim$ to indicate a correspondence. This series expansion of $F(x)$ is not convergent for any fixed value of x: the ratio test gives

$$\frac{u_n}{u_{n-1}} = \frac{n!}{x^{n+1}} \cdot \frac{x^n}{(n-1)!} = \frac{n}{x}$$ (6)

so that the limit for $n \to \infty$ does not exist. However, even though the series (5) is divergent, it represents $F(x)$ in some sense.

Since the terms of the series at first decrease in magnitude and then eventually increase to infinity, it would seem reasonable that the optimum approximation to $F(x)$, for a given x, would be the sum of those terms that are decreasing. This can be shown as follows: since

$$\frac{1 - (-1)^N t^N}{1 + t} = 1 - t + t^2 - \cdots + (-1)^{N-1} t^{N-1},$$ (7)

then, for all t,

$$\frac{1}{1 + t} = \sum_0^{N-1} (-1)^n t^n + \frac{(-1)^N t^N}{1 + t}.$$ (8)

Using this expansion in the integral for $F(x)$ we obtain

$$F(x) = \sum_{n=0}^{N-1} \frac{(-1)^n n!}{x^{n+1}} + (-1)^N \int_0^\infty \frac{e^{-xt} t^N}{1 + t} \, dt,$$ (9)

where the first N terms of (9) coincide with the first N terms of (5). Therefore the error in approximating $F(x)$ by the sum of the first N terms is

$$E_N(x) = (-1)^N \int_0^\infty \frac{e^{-xt} t^N}{1 + t} \, dt.$$ (10)

Since

$$|E_N| = \int_0^\infty \frac{e^{-xt} t^N}{1 + t} \, dt < \int_0^\infty \frac{e^{-xt} t^N}{1} \, dt = \frac{N!}{x^{N+1}},$$ (11)

where the inequality follows on replacing $(1 + t)^{-1}$ by its maximum value 1, we see that the error in the approximation

$$F(x) \simeq \sum_{n=0}^{N-1} \frac{(-1)^n n!}{x^{n+1}} \qquad (12)$$

of (5) is always less than the first term neglected. (Note that this last statement is not true for all cases we consider.)

The utility of the preceding approximation can be illustrated by the following numerical examples: consider

$$F(2) \sim \frac{1}{2} - \frac{1}{4} + \frac{2!}{8} - \frac{3!}{16} + \frac{4!}{32} - \cdots = \frac{1}{2} - \frac{1}{4} + \frac{1}{4} - \frac{3}{8} + \frac{3}{4} - \cdots$$

The approximation to $F(2)$ is obviously poor, and the best value is obtained from the first term $(1/2)$ which gives an error of fity percent $(1/4)$. On the other hand, for $F(10)$, we find

$$F(10) \sim \frac{1}{10} - \frac{1}{10^2} + \frac{2}{10^3} - \frac{3!}{10^4} + \cdots = 0.092 - \frac{6}{10^4} + \cdots ,$$

so that the first three terms suffice to within an error of 0.0006.

The above gives an example of an asymptotic expansion and a demonstration of its limitations. Finding asymptotic expansions for such functions as $F(x)$ is largely an art, but there exist a few standard methods.

DEFINITION OF ASYMPTOTIC EXPANSION

Before continuing further, we introduce some definitions and notations.

Definition 1.1: We say that

$$g(x) = O[\phi(x)] \qquad (13a)$$

if there exists a constant A and a number x_0 such that

$$|g(x)| < A |\phi(x)| \qquad (13b)$$

whenever $x > x_0$. For example, we have proved that

$$\int_0^\infty \frac{e^{-xt}}{1+t} dt - \sum_{n=0}^{N-1} \frac{(-1)^n n!}{x^{n+1}} = E_N(x) = O\left(\frac{1}{x^{N+1}}\right), \qquad (14)$$

because by (11),

$$|E_N(x)| < \frac{N!}{x^{N+1}} = \frac{A}{x^{N+1}} \qquad (15)$$

if we choose $A = N!$.

Definition 1.2: We say that

$$g(x) = o[\phi(x)] \tag{16a}$$

if

$$\lim_{x \to \infty} \frac{g(x)}{\phi(x)} = 0. \tag{16b}$$

For example, from (15) we have

$$E_N(x) = o\left(\frac{1}{x^N}\right)$$

because

$$x^N |E_N(x)| < \frac{N! x^N}{x^{N+1}} \to 0 \quad \text{as } x \to \infty.$$

Definition 1.3: Suppose that for $N = 1, 2, \ldots$, we have a sequence

$$f(x) = c_0 \phi_0(x) + c_1 \phi_1(x) + \cdots + c_{N-1} \phi_{N-1}(x) + O[\phi_N(x)]$$

where, for all N,

$$\phi_N(x) = o[\phi_{N-1}(x)].$$

We mean a sequence of improvements for $x \to \infty$:

$$f(x) = c_0 \phi_0 + O[\phi_1], \quad \phi_1 = o[\phi_0];$$

$$f(x) = c_0 \phi_0 + c_1 \phi_1 + O[\phi_2], \quad \phi_2 = o[\phi_1]; \text{ etc}$$

Then we write

$$f(x) \sim c_0 \phi_0 + c_1 \phi_1 + \cdots + c_{N-1} \phi_{N-1} + \cdots \tag{17}$$

and call the right-hand side the asymptotic expansion of $f(x)$, an idea introduced by Poincaré.

It is important to recognize that an asymptotic series need not converge, but a convergent series is necessarily asymptotic. For example consider a function $\psi(t)$ expressed as an infinite series:

$$\psi(t) = \sum_0^\infty a_n t^n. \tag{18}$$

Assume that the series for $\psi(t)$ converges. Then

$$\psi(t) = \sum_{n=0}^{N-1} a_n t^n + O(t^N) \tag{19}$$

and $t^N = o(t^{N-1})$ as $t \to 0$. Thus a convergent series satisfies definition 1.3. (For convergence, we talk about what happens at a given x as $n \to \infty$, but when we say asymptotic, we consider what happens for finite n as x varies.)

If we have an asymptotic expression for $f(x)$, then it is unique in terms of a given set ϕ_n. This is shown by considering two separate asymptotic series expansions for $f(x)$, say

$$f(x) \sim \sum_i c_i \phi_i(x) \tag{20}$$

and

$$f(x) \sim \sum_i d_i \phi_i(x). \tag{21}$$

Subtracting (21) from (20), we obtain

$$f(x) - f(x) \sim \sum_i (c_i - d_i) \phi_i \sim 0.$$

Assume that c_5, say, is the first of the c's that is different from the corresponding d; then

$$0 \sim (c_5 - d_5) \phi_5 + O[\phi_6]. \tag{22}$$

Dividing by $c_5 - d_5 \neq 0$, we obtain

$$0 \sim \phi_5 + \frac{1}{c_5 - d_5} O[\phi_6],$$

which means that

$$\left|0 - \phi_5\right| < \frac{A}{c_5 - d_5} \phi_6,$$

or equivalently,

$$\left|\frac{\phi_5}{\phi_6}\right| < \frac{A}{c_5 - d_5}. \tag{23}$$

However, by assumption, $\varphi_6 = o[\varphi_5]$, and the left-hand side becomes infinite as x grows large; therefore there does not exist a constant A satisfying (23). This contradiction shows that $c_5 - d_5$ must equal zero.

We show that some functions have zero asymptotic expansions. First, note that if $f(x) \sim \sum c_n/x^n$, then

$$c_0 = \lim_{x \to \infty} f(x) \tag{24}$$

$$c_1 = \lim_{x \to \infty} x[f(x) - c_0]$$

$$c_2 = \lim_{x \to \infty} x^2\left[f(x) - c_0 - \frac{c_1}{x}\right].$$

Put $f(x) = e^{-x} \sim \sum c_n/x^n$; then we see that $c_0 = 0$, and then $c_1 = 0$, etc., so that

$$e^{-x} \sim 0 + \frac{0}{x} + \frac{0}{x^2} + \cdots . \qquad (25)$$

This example shows that two different functions, for example, $f(x)$ and $f(x) + e^{-x}$, may have the same asympotic expansion.

OPERATIONS WITH ASYMPTOTIC SERIES

Suppose

$$f(x) \sim \sum_n \frac{a_n}{x^n} ,$$

$$(26)$$

$$g(x) \sim \sum_n \frac{b_n}{x^n} .$$

Then the usual rules hold for addition,

$$f(x) + g(x) \sim (a_0 + b_0) + \frac{a_1 + b_1}{x} + \cdots ,$$

and for multiplication,

$$f(x)g(x) \sim a_0 b_0 + \frac{(a_0 b_1 + a_1 b_0)}{x} + \cdots .$$

We also have the composite form

$$f[g(x)] \sim \sum_n a_n \left(\sum_m \frac{b_m}{x^m} \right)^{-n} , \qquad (27)$$

which says that the asymptotic expansion for $f[g(x)]$ can be obtained by re-placing x in the asymptotic expansion for $f(x)$ by the asymptotic expansion for g.

The asymptotic expansion can be integrated unconditionally. Suppose

$$f(x) = \sum_{n=2}^{N-1} c_n x^{-n} + O(x^{-N}) , \qquad (28)$$

then

$$f(x) - \sum_{n=2}^{N-1} c_n x^{-n} = O(x^{-N}) ,$$

and

$$\left| f(x) - \sum_{n=2}^{N-1} c_n x^{-n} \right| < A \left| x^{-N} \right| ,$$

where A is some finite number. We integrate term by term,

$$\left| \int_{x_1}^{\infty} \left\{ f(x) - \sum_{n=2}^{N-1} c_n x^{-n} \right\} dx \right| < A \int_{x_1}^{\infty} \frac{dx}{x^N} = \frac{A}{(N-1)x_1^{N-1}}, \qquad (29)$$

and rewrite to get

$$\left| x^{N-1} \left\{ \int_{x_1}^{\infty} f(x) \ dx - \int_{x_1}^{\infty} \sum_{n=2}^{N-1} c_n x^{-n} \ dx \right\} \right| < \frac{A}{N-1}. \qquad (30)$$

Thus by the definition of asymptotic series expansions, it is seen that the term-by-term integration of $\sum_{n=2}^{N-1} c_n x^{-n}$ gives an asymptotic expansion for $\int_{x_1}^{\infty} f(x)\,dx$. The terms for $n = 0$ and $n = 1$ are omitted because they give divergent integrals at infinity.

Differentiation of asymptotic expansions is not always possible. For example,

$$e^{-x} \sin(e^x) \sim 0 + \frac{0}{x} + \frac{0}{x^2} + \cdots, \qquad x \to \infty,$$

but the derivative

$$\cos(e^x) - e^{-x} \sin(e^x)$$

does not have an asymptotic expansion. However, it can be proved that if both $f(x)$ and $f'(x)$ have asymptotic expansions, then the asymptotic expansion for $f'(x)$ is obtained by termwise differentiation of the asymptotic expansion for $f(x)$.

INTEGRATION BY PARTS

An alternative method of obtaining an asymptotic expansion of an integral such as $\int_0^{\infty} [e^{-xt}/(1 + t)]\,dt = F(x)$ of (1) is successive integration by parts. Thus if we integrate $F(x)$ by parts we obtain

$$F(x) = -\frac{e^{-xt}}{x(1 + t)} \Big|_0^{\infty} - \frac{1}{x} \int_0^{\infty} \frac{e^{-xt}}{(1 + t)^2} \ dt = \frac{1}{x} - \frac{1}{x} \int_0^{\infty} \frac{e^{-xt}}{(1 + t)^2} \ dt.$$

Integrating by parts N times in succession, we get

$$F(x) = \frac{1}{x} \sum_{n=0}^{N-1} \frac{n!}{(-x)^n} + \frac{N!}{(-x)^N} \int_0^{\infty} \frac{e^{-xt}}{(1 + t)^N} \ dt.$$

As t goes from 0 to ∞, the function $(1 + t)^{-N}$ decreases steadily from 1 to 0; consequently, $\int_0^{\infty} \frac{e^{-xt}}{(1 + t)^N} \ dt < \int_0^{\infty} e^{-xt} \ dt = \frac{1}{x}$, and

$$\left| F(x) - \frac{1}{x} \sum_0^{N-1} \frac{n!}{(-x)^n} \right| < \frac{N!}{x^{N+1}}$$

just as we obtained before.

Similarly, for the integral $\int_x^\infty (e^{-t}/t)\,dt$, we integrate by parts to get

$$\int_x^\infty \frac{e^{-t}}{t}\,dt = \frac{e^{-x}}{x} - \int_x^\infty \frac{e^{-t}}{t^2}\,dt \qquad (31)$$

$$= \frac{e^{-x}}{x} - \frac{e^{-x}}{x^2} + \int_x^\infty \frac{2e^{-t}}{t^3}\,dt = e^{-x}\left(\frac{1}{x} - \frac{1!}{x^2} + \frac{2!}{x^3} - \frac{3!}{x^4} + \cdots\right),$$

and since

$$\left|\int_x^\infty \frac{e^{-t}}{t^n}\,dt\right| < \frac{1}{x^n}\int_x^\infty e^{-t}\,dt = \frac{e^{-x}}{x^n}, \qquad (32)$$

we see that the quantity $e^{-x}(n-1)!/x^n$ is a bound for the error.

As another example, consider the integral $\int_0^x e^{-t^2}\,dt$. We have

$$G(x) = \int_0^x e^{-t^2}\,dt = \int_0^x \left(1 - t^2 + \frac{t^4}{2!} - \cdots\right)dt \qquad (33)$$

$$= x - \frac{x^3}{3} + \cdots,$$

and although this series holds for all x, it is slowly convergent for $x \geq 3$. Thus an asymptotic expansion is useful for large values of x. Since $\lim_{x\to\infty} G(x) = \int_0^\infty e^{-t^2}\,dt = \frac{1}{2}\sqrt{\pi}$, we may write

$$G(x) = \int_0^\infty e^{-t^2}\,dt - \int_0^\infty e^{-t^2}\,dt = \frac{\sqrt{\pi}}{2} - \int_x^\infty e^{-t^2}\,dt. \qquad (34)$$

Now, by integration by parts,

$$\int_x^\infty e^{-t^2}\,dt = \int_x^\infty te^{-t^2}\frac{1}{t}\,dt = -\left.\frac{e^{-t^2}}{2t}\right|_x^\infty - \int_x^\infty \frac{e^{-t^2}}{2t^2}\,dt,$$

so

$$G(x) = \frac{\sqrt{\pi}}{2} - \frac{e^{-x^2}}{2x} + \int_x^\infty \frac{e^{-t^2}}{2t^2}\,dt.$$

Further integration by parts gives the expansion

$$G(x) \sim \frac{\sqrt{\pi}}{2} - \frac{e^{-x^2}}{2x}\sum_0^\infty (-1)^n \binom{-1/2}{n}\frac{n!}{x^{2n}},$$

where $\binom{-1/2}{n}$ is the binomial coefficient. The final form follows more directly by introducing $s = t^2$ and expanding $\int_x^\infty e^{-t^2}\,dt = \int_{x^2}^\infty \frac{e^{-s}}{2s^{1/2}}\,ds$.

As a final illustration of this method, we apply it to Fresnel's integral. We have

$$F(x) = \int_0^x e^{it^2} dt = \left[\int_0^\infty - \int_x^\infty\right] e^{it^2} dt$$

$$= \frac{1}{2}\sqrt{\pi}\, e^{i\pi/4} - \int_x^\infty e^{it^2} dt = F(\infty) + E(x), \tag{35}$$

where we used the well-known result for $F(\infty)$. We now change variables and use integration by parts to get

$$2E(x) = -\int_{x^2}^\infty \frac{e^{is}}{\sqrt{s}}\, ds = \frac{e^{is}}{i\sqrt{s}}\Big|_{x^2}^\infty - \frac{1}{2i}\int_{x^2}^\infty \frac{e^{is}}{s^{3/2}}\, ds$$

$$= \frac{e^{ix^2}}{ix}\sum_0^{N-1}\frac{C_n}{x^{2n}} - C_N\int_{x^2}^\infty \frac{e^{is}}{s^{N+1/2}}\, ds, \tag{36}$$

where $C_n = 1\cdot 3\cdot 5\cdots(2n-1)/(i\,2)^n$. If we try to estimate the error term as we have done before, namely,

$$\left|\int_{x^2}^\infty \frac{e^{is}}{s^{N+1/2}}\, ds\right| < \frac{1}{x^{2N+1}}\int_{x^2}^\infty |e^{is}|\, ds, \tag{37}$$

we find that the last integral does not converge. This difficulty can be avoided by leaving some power of x in the denominator of the integrand:

$$\left|\int_{x^2}^\infty \frac{e^{is}}{s^{N+1/2}}\, ds\right| < \frac{1}{x^{2N-2}}\int_{x^2}^\infty \frac{|e^{is}|ds}{s^{3/2}} = \frac{1}{x^{2N-2}}\frac{2}{x} = O\left(\frac{1}{x^{2N-1}}\right). \tag{38}$$

This estimate is not too good since it is not of the order of the next term.

To get a better estimate, we integrate the remainder in (36) by parts to obtain

$$I = \int_{x^2}^\infty \frac{e^{is}}{s^{N+1/2}}\, ds = \frac{e^{is}}{is^{N+1/2}}\Big|_{x^2}^\infty + \frac{(N+1/2)}{i}\int_{x^2}^\infty \frac{e^{is}}{s^{N+3/2}}\, ds, \tag{39}$$

and now we replace e^{is} in the integral by unity; thus

$$I = \frac{ie^{ix^2}}{x^{2N+1}} - i\left(N+\frac{1}{2}\right)O\left(\frac{1}{x^{2N+1}}\right). \tag{40}$$

Since each term is $O(x^{-2N-1})$, we have thus shown that (36) is an asymptotic expansion.

IMPROVING AN EXPANSION

We saw earlier that the asymptotic expansion of the function

$$F(x) = \int_0^\infty \frac{e^{-xt}}{1+t}\, dt = \frac{1}{x} - \frac{1}{x^2} + \frac{2!}{x^3} - \frac{3!}{x^4} + \cdots. \tag{41}$$

for small or moderate ($x \leq 2$) values of the variable is very poor. We now construct another expansion more suitable for small x. We observe that the first two terms of the asymptotic series for large x may be combined as follows:

$$F(x) \sim \frac{1}{x} - \frac{1}{x^2} = \frac{1}{x}\left(1 - \frac{1}{x}\right) \approx \frac{1}{x}\frac{1}{(1 + 1/x)} = \frac{1}{1 + x}. \tag{42}$$

This suggests that $F(x) \sim 1/(x + 1)$, which is valid for very large x, may also be useful for small x. In any case we can estimate the error. We have

$$E = \int_0^\infty \frac{e^{-xt}}{1 + t}\, dt - \frac{1}{x + 1}, \tag{43}$$

and if we multiply the second term by $x \int_0^\infty e^{-xt}dt = 1$, we may write

$$E = \int_0^\infty e^{-xt}\left[\frac{1}{1 + t} - \frac{1}{1 + 1/x}\right] dt. \tag{44}$$

The bracket under the last integral changes sign between zero and infinity; this causes difficulty in estimating the value of the integral in (44). We therefore rearrange the terms to obtain

$$E = \int_0^\infty \left\{ e^{-xt}\left[\frac{1}{x} - t\right]\right\} \frac{dt}{(1 + t)(1 + 1/x)}, \tag{45}$$

in order to exploit the fact that the function in braces has an exact integral, namely,

$$\int e^{-xt}\left[\frac{1}{x} - t\right] dt = \frac{te^{-xt}}{x}. \tag{46}$$

Integration by parts gives

$$E = \frac{1}{x}\left. \frac{e^{-xt}\, t}{(1 + t)(1 + 1/x)}\right|_0^\infty + \frac{1}{x}\frac{1}{(1 + 1/x)}\int_0^\infty \frac{t\, e^{-xt}\, dt}{(1 + t)^2} \tag{47}$$

$$= \frac{1}{1 + x}\int_0^\infty \frac{t\, e^{-xt}\, dt}{(1 + t)^2},$$

and since the final integrand is positive and monotonic, we obtain directly

$$E < \frac{1}{1 + x}\int_0^\infty t\, e^{-xt}\, dt = \frac{1}{x^2(1 + x)}. \tag{48}$$

In particular, for $x = 2$, we have from (42) and (48) that $F(2) \sim 1/3$ with an error less than $1/12$. The tabulated value of $F(2)$ is 0.3613 and the actual error is 0.3613-0.3333 = 0.028. Thus we see that $1/(x + 1)$ is a closer approximation to $F(x)$ than either $1/x$ or $1/x - 1/x^2$.

3.2 WATSON'S LEMMA

We have developed asymptotic expansions for an integral of the form

$F(x) = \int_0^\infty e^{-xt} f(t)\, dt$ (a Laplace integral) by two methods; integration by parts, and term-by-term integration of the series expansion for $f(t)$. The justification for the second method is given by Watson's lemma, which proves that an asymptotic expansion exists.

We assume (*i*) that $f(t)$ has a power series expansion which converges for $|t| < R$; this series determines the behavior of $f(t)$ for small values of t. For large values of t we assume (*ii*) that there exists an $\alpha > 0$ such that $f(t) = O(e^{\alpha t})$. Then if we replace $f(t)$ by its Taylor series and integrate term-by-term, we get

$$\int_0^\infty e^{-xt} f(t)\, dt = \int_0^\infty e^{-xt} \sum_0^\infty \frac{f^{(K)}(0) t^K}{K!} \, dt \sim \sum_{K=0}^\infty \frac{f^{(K)}(0)}{x^{K+1}}. \tag{49}$$

In general, this series does not converge, but the following argument shows it is always asymptotic; that is,

$$\left| \int_0^\infty e^{-xt} f(t)\, dt - \sum_{K=0}^{N-1} \frac{f^{(K)}(0)}{x^{K+1}} \right| = O\left(\frac{1}{x^{N+1}} \right), \tag{50}$$

which is known as Watson's lemma.

From assumption (*i*), we have

$$\left| f(t) - \sum_0^{N-1} \frac{f^{(K)}(0) t^K}{K!} \right| = \left| \frac{f^{(N)}(0)}{N!} t^N + \cdots \right|$$

$$= \left| t^N \sum_{k=0}^\infty \frac{f^{(N+k)}(0)}{(N+k)!} t^k \right| = O(t^N); \tag{51}$$

that is, because the power series for $f(t)$ converges, the sum over k is bounded for $t < R/2$, say. Next, we modify (51) so that it will be valid for all values of t. Since the series $\sum_0^{N-1}(f^{(k)}(0)/k!)t^k$ is a polynominal and therefore $O(e^{\alpha t})$ for all values of t, it follows from assumption (*ii*) that

$$\left| f(t) - \sum_0^{N-1} \frac{f^{(k)}(0) t^k}{k!} \right| = O(e^{\alpha t}) \tag{52}$$

for all t. However, we can do better with the help of (51); since $t^N = O(t^N e^{\alpha t})$ for all values of t, and $e^{\alpha t} = O(t^N e^{\alpha t})$ for all sufficiently large t, we conclude that

$$\left| f(t) - \sum_0^{N-1} \frac{f^{(k)}(0) t^k}{k!} \right| = O(t^N e^{\alpha t}) \tag{53}$$

for all t. Then

$$\int_0^\infty e^{-xt} \, f(t) \, dt \; - \; \sum_0^{N-1} \frac{f^{(K)}(0)}{x^{K+1}} \; = \; \int_0^\infty e^{-xt} \left[f(t) \, - \, \sum_0^{N-1} \frac{f^{(K)}(0) \, t^K}{K!} \right] dt$$

$$= O\left[\int e^{-xt} \, t^N \, e^{\alpha t} \, dt \right] \; = \; O\left[\int e^{-(x-\alpha)t} \, t^N \, dt \right]$$

$$= O\left(\frac{1}{(x \, - \, \alpha)^{N+1}} \right) \; = \; O\left(\frac{1}{x^{N+1}} \right) \tag{54}$$

for sufficiently large x. Thus we have proved Watson's lemma (50), that is, that the series in (49) is asymptotic.

We shall apply (49) to $F(x)$ of (1) and consider variations of the original asymptotic series (5). From (5), when $x = 4$, we find

$$F(4) = \frac{1}{4} - \frac{1!}{4^2} + \frac{2!}{4^3} - \frac{3!}{4^4} + \frac{4!}{4^5} - \frac{5!}{4^6} + \cdots$$

or

$$4F(4) = 1 - \frac{1}{4} + \frac{2!}{4^2} - \frac{3!}{4^3} + \frac{4!}{4^4} - \frac{5!}{4^5} + \cdots \equiv \sum_N .$$

Now,

$$\sum_4 - 0.78125, \qquad \sum_5 = 0.87500 .$$

Upon comparison with the tabulated value $4F(4) = 0.82533$, it is seen that $\sum_4$ and $\sum_5$ are fairly close. Since $\sum_4$ is too small and $\sum_5$ is too large, it seems reasonable that their average value will be more accurate. Averaging $\sum_4$ and $\sum_5$, we obtain $(\sum_4 + \sum_5)/2 = 0.82813$, which has an accuracy of 0.3%.

We can improve on the accuracy obtained by averaging if we approximate the remainder. Thus we consider the remainder after the fourth term in the previous example; since

$$\frac{1}{1 + t} = 1 - t + t^2 - t^3 - \frac{t^4}{1 + t} ,$$

the error in the integral will be

$$R = \int_0^\infty e^{-xt} \, \frac{t^4}{1 + t} \, dt . \tag{55}$$

This expression for the error R is of little use since it is not in a form suitable for the application of Watson's lemma; that is, the approximation

$$\int_0^\infty e^{-xt} \, f(t) \, dt \sim f(0) \int_0^\infty e^{-xt} \, dt, \qquad f(t) = \frac{t^4}{1 + t}$$

gives zero for the value of the integral. To obtain a useful form of R, we consider the integral

$$I = \int_0^\infty e^{-4t} t^4 f(t)\, dt$$

and integrate by parts in such a way that we obtain again an integral of the same form, namely,

$$\int_0^\infty e^{-4t} t^4 g(t)\, dt .$$

The fact that

$$\frac{d}{dt}\left(e^{-4t}\right) = 4e^{-4t} t^3(1 - t) ,$$

suggests writing the original integral as

$$I = \int_0^\infty e^{-4t} t^3 \left[t f(t)\right] dt = f(1) \int_0^\infty e^{-4t} t^3\, dt$$

$$+ \int_0^\infty e^{-4t} t^3 \left[t f(t) - f(1)\right] dt = I_1 + I_2 .$$

Since the quantity in brackets in the last integral I_2 vanishes when $t = 1$, we may write

$$I_2 = \int_0^\infty e^{-4t} t^3 (t - 1) \left[\frac{t f(t) - f(1)}{t - 1}\right] dt ,$$

and integrate by parts to obtain

$$I_2 = \frac{e^{-4t} t^4}{-4}\left[\frac{t f(t) - f(1)}{t - 1}\right]\Bigg|_0^\infty + \frac{1}{4}\int_0^\infty e^{-4t} t^4 g(t)\, dt ,$$

where the first term vanishes, and where

$$g(t) = \frac{d}{dt}\left[\frac{t f(t) - f(1)}{t - 1}\right] .$$

Thus, using this value for I_2, and with $I_1 = 3!\, f(1)/4^4$, we have

$$\int_0^\infty e^{-4t} t^4 f(t)\, dt = \frac{3!}{4^4} f(1) + \frac{1}{4}\int_0^\infty e^{-4t} t^4 g(t)\, dt . \qquad (56)$$

Let us apply (56) to (55). We obtain

$$R = \frac{1}{2}\cdot\frac{3!}{4^4} + \frac{1}{4}\int_0^\infty e^{-4t} t^4 g(t)\, dt = R_1 + R_2(g) ,$$

where

$$g(t) = \frac{d}{dt}\left[2(1 + t)^{-1}\right] = -2(1 + t)^{-2} .$$

Applying (56) again, to $R_2(g)$, we get

$$R = \frac{1}{2}\frac{3!}{4^4} - \frac{1}{32}\frac{3!}{4^4} + \frac{1}{16}\int_0^\infty e^{-4t}\, t^4\, h(t)\, dt = R_1 + S_1 + S_2(h),$$

where

$$h(t) = \frac{d}{dt}\left[\frac{t\,g(t) - g(1)}{t - 1}\right].$$

If we neglect $S_2(h)$, we have

$$R = \frac{15}{32}\frac{3!}{4^4},$$

and using this to calculate $4F(4)$, we find $4F(4) = 0.82508$, with an error of 0.03%.

We can also use Watson's lemma to obtain asymptotic expansions of integrals such as

$$\int_{-\infty}^\infty e^{-xt^2}\, f(t)\, dt = \left[\int_0^\infty + \int_{-\infty}^0\right] e^{-xt^2}\, f(t)\, dt \tag{57}$$

if we change variables by setting $t^2 = s$. Of course, $f(s^{\frac{1}{2}})$ may not be analytic, but it is enough that there exists a convergent expansion of the form

$$f(t) = t^\beta \sum_0^\infty c_K\, t^K, \tag{58}$$

and that also $f(t) = O(e^{\beta t^2})$ in order for the above method to apply.

STIRLING'S FORMULA

The following example is not of the previously considered type; however, it can be handled by a suitable change of variable.

We study the gamma function integral $\Gamma(n + 1) = \Gamma$,

$$\Gamma = \int_0^\infty e^{-t}\, t^n\, dt, \tag{59}$$

whose value is known to be $n!$. We seek to determine how $n!$ behaves for a large value of n, say $n = 1000$. Instead of proceeding directly to an appropriate form, we first illustrate the tentative considerations that are often called for.

To use Watson's lemma, it is desirable that the integrand have the parameter in the exponent. Thus we first rewrite (59) in the form

$$\int_0^\infty e^{-t + n\,\ln t}\, dt. \tag{60}$$

Since Watson's lemma can still not be applied, we try another change of variables, $t - n\ln t = u$, $dt\,[1 - n/t] = du$, and obtain

$$\int_0^\infty e^{-t+n \ln t} \, dt = \int_\infty^\infty \epsilon^{-u} \frac{du}{1 - \dfrac{n}{t}} . \tag{61}$$

We see that $u = \infty$ when $t = \infty$ or 0, and that for $t = n$ the integral tends to infinity; this is not a one-to-one change of variables, because t is not a single-valued function of u, and special difficulty arises when $t = n$.

We then try a different substitution in (60), $t = n(1 + \tau)$, and obtain

$$\Gamma = e^{-n} e^{n \ln n} \int_{-1}^\infty e^{-n\tau + n \ln (1+\tau)} n \, d\tau$$

$$= e^{-n} n^{n+1} \int_{-1}^\infty e^{-n[\tau - \ln(1+\tau)]} \, d\tau . \tag{62}$$

This last integral would be in the proper form for the application of Watson's lemma if we made the substitution $[\tau - \ln(1 + \tau)] = \sigma$. But again this substitution leads to difficulty since τ is not a single-valued function of σ. The graph of σ as a function of τ looks like Figure 3.1.

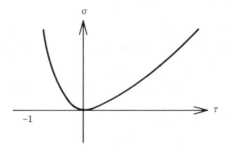

Figure 3.1

This curve has the appearance of a parabola and suggests putting $\sigma = s^2$ in order to get a one-to-one transformation. Put $[\tau - \ln(1 + \tau)] = s^2$; then $\tau \, d\tau / (1+\tau) = 2s \, ds$

and

$$\Gamma = e^{-n} n^{n+1} \int_{-\infty}^\infty e^{-ns^2} \frac{1 + \tau}{\tau} 2s \, ds \tag{63}$$

$$= e^{-n} n^{n+1} \left[2 \int_{-\infty}^\infty \frac{e^{-ns^2} s}{\tau} \, ds + 2 \int_{-\infty}^\infty e^{-ns^2} s \, ds \right].$$

The last integral in the bracket is zero. To evaluate the remaining integral we expand $s^2 = \tau - \ln(1 + \tau)$ by using the power series for $\log(1 + \tau)$:

$$s^2 = \tau - \left[\tau - \frac{\tau^2}{2} + \frac{\tau^3}{3} - \cdots \right] = \frac{\tau^2}{2} - \frac{\tau^3}{3} + \cdots ,$$

and

$$s = \frac{\tau}{\sqrt{2}} \left[1 - \frac{2\tau}{3} + \cdots\right]^{1/2} .$$

Inverting the series, we obtain

$$\tau = \sqrt{2}\, s \left[1 - \frac{2}{3}\sqrt{2}\, s + \cdots\right]^{-1/2} ,$$

and the binomial expansion converges near $s = 0$. Using the first term $s = \tau/\sqrt{2}$,

$$2 \int_{-\infty}^{\infty} e^{-ns^2}\, \frac{s}{\tau}\, ds \approx \sqrt{2} \int_{-\infty}^{\infty} e^{-ns^2}\, ds = \sqrt{\frac{2\pi}{n}}, \tag{64}$$

and substituting into (63), we obtain Stirling's formula

$$n! \sim e^{-n} n^{n+1/2} \sqrt{2\pi} = n^n e^{-n} \sqrt{2\pi n}. \tag{65}$$

We shall now give an estimate of the error made by using the approximation (65). This can be done by finding the error in the approximation (64). We have

$$s^2 = \tau - \ln(1 + \tau) = \int_0^\tau \frac{\xi\, d\xi}{1 + \xi} = \int_0^\tau \xi\, d\xi - \int_0^\tau \frac{\xi^2}{1 + \xi}\, d\xi \tag{66}$$

$$= \frac{\tau^2}{2} - \frac{\tau^3}{2}\, E ,$$

Since for $-\frac{1}{2} \leq \tau$, the function $(1 + \tau)^{-1} \leq 2$, we find that

$$\left|\frac{\tau^3 E}{2}\right| = \left|\int_0^\tau \frac{\xi^2}{1 + \xi}\, d\xi\right| \leq 2\left|\int_0^\tau \xi^2\, d\xi\right| \leq \frac{2|\tau^3|}{3} ,$$

so that $|E| < 4/3$. From (66), we find that

$$s = \frac{\tau}{\sqrt{2}}\, (1 - \tau E)^{1/2} , \tag{67}$$

and then

$$\left|\frac{s}{\tau} - \frac{1}{\sqrt{2}}\right| = \left|(1 - \tau E)^{1/2} - 1\right|\frac{1}{\sqrt{2}}$$

$$= \left|\frac{\tau E}{(1 - \tau E)^{1/2} + 1}\right|\frac{1}{\sqrt{2}} \leq \frac{4}{3}|\tau| , \tag{68}$$

for $\tau \geq -\frac{1}{2}$. But from (67)

$$|\tau| = \left|\frac{2^{1/2} s}{(1 - \tau E)^{1/2}}\right| \leq 2^{1/2}\, |s| < 2\, |s| ,$$

so that (68) implies

$$\left| \frac{s}{\tau} \cdot 2^{-1/2} \right| \le \frac{8}{3} |s| . \tag{69}$$

For $-1 \le \tau \le -\frac{1}{2}$, we have

$$\left| \frac{s}{\tau} - 2^{-1/2} \right| \le 2 |s| + 2^{-1/2} \le 4 |s| . \tag{70}$$

Combining (69) and (70), we see that

$$\left| \frac{s}{\tau} - 2^{-1/2} \right| \le 4 |s| \tag{71}$$

for all values of s. Using this in the integral required for (63), we find instead of (64) that

$$2 \int_{-\infty}^{\infty} e^{-ns} \frac{s}{\tau} ds = \sqrt{2} \int e^{-ns^2} ds + E_1 = \left(\frac{2\pi}{n} \right)^{1/2} + E_1,$$

with $E_1 = 2 \int e^{-ns^2} \left[\frac{s}{\tau} - \frac{1}{\sqrt{2}} \right] ds$. Using (71), we find that

$$\left| E_1 \right| \le 8 \int_{-\infty}^{\infty} e^{-ns^2} |s| ds = \frac{8}{n},$$

so that we have reduced (63) to

$$n! = (2\pi n)^{1/2} \, n^n \, e^{-n} (1 + E_2), \tag{72}$$

where

$$|E_2| < \frac{8(2\pi)^{-1/2}}{n^{1/2}} .$$

LAPLACE'S METHOD

A typical class of integrals for which we can obtain asymptotic expansions is

$$I = \int_0^{\infty} e^{-xf(t)} g(t) \, dt , \tag{73}$$

where

$$f(t) \to \infty \text{ as } t \to \infty .$$

We rewrite (73) in the form of a Laplace integral by setting $f(t) = u$ and $dt = du/f'(t)$. Then (73) becomes

$$I = \int e^{-xu} \left(\frac{g(t)}{f'(t)} \right) du = \int e^{-xu} \, h(u) \, du \,, \tag{74}$$

where $h(u) = g(t)/f'(t)$. Watson's lemma can be applied to (74) except if $g(t)/f'(t)$ becomes infinite inside the interval of integration. The integrand becomes singular at a stationary point of $f(t)$, that is, at a point $t = t_0$ such that $f'(t_0) = 0$. In the neighborhood of t_0, we may in general approximate $f(t)$ by a parabola. This follows from

$$f(t) = f(t_0) + (t - t_0) f'(t_0) + \frac{(t - t_0)^2}{2} f''(t_0) + \cdots$$

$$\tag{75}$$

$$= f(t_0) + \frac{(t - t_0)^2}{2} f''(t_0) + \cdots \equiv f(t_0) + s^2 \,,$$

where

$$s \simeq (t - t_0) \sqrt{\frac{f''(t_0)}{2}} \text{ for } f''(t_0) \neq 0. \tag{76}$$

Using $f(t) = f(t_0) + s^2$ in (73), we obtain

$$I = \int_0^\infty e^{-x f(t)} \, g(t) \, dt = e^{-x f(t_0)} \int e^{-xs^2} \frac{g(t)}{f'(t)} \, 2s \, ds \,. \tag{77}$$

To evaluate this integral asymptotically, we need the behavior of $[g(t)/f'(t)] \, 2s$ for $s = 0$. Using (76), and the corresponding expansion of $f'(t)$ around t_0,

$$f'(t) = f'(t_0) + (t - t_0) f''(t_0) + \cdots = (t - t_0) f''(t_0) + \cdots \,, \tag{78}$$

we have

$$\frac{s}{f'(t)} \approx (t - t_0) \sqrt{\frac{f''(t_0)}{2}} \, \frac{1}{(t - t_0) f''(t_0)} = \frac{1}{\sqrt{2 f''(t_0)}} \,. \tag{79}$$

Consequently, from (79) and from $g(t) \approx g(t_0)$, we reduce (77) to

$$I = 2 e^{-x f(t_0)} \, \frac{g(t_0)}{\sqrt{2 f''(t_0)}} \int_{-a}^\infty e^{-xs^2} \, ds \,, \quad a = t_0 \sqrt{f''(t_0)/2} \,. \tag{80}$$

We have

$$\int_{-a}^\infty e^{-xs^2} \, ds = \int_{-\infty}^\infty e^{-xs^2} \, ds - \int_{-\infty}^{-a} e^{-xs^2} \, ds \sim \sqrt{\frac{\pi}{x}} - \frac{e^{-xa^2}}{2ax} + \cdots \tag{81}$$

where the term $e^{-xa^2}/2ax$ is much smaller than the first term (since $a \neq 0$). Thus as long as t_0 is in the range of integration we keep only the first term and use

$$I \approx 2 e^{-x f(t_0)} \, \frac{g(t_0)}{\sqrt{2 f''(t_0)}} \int_{-\infty}^\infty e^{-xs^2} \, ds \approx \sqrt{\frac{2\pi}{x f''(t_0)}} \, e^{-x f(t_0)} \, g(t_0) \,. \tag{82}$$

3.3 METHOD OF STEEPEST DESCENT

An extension of Laplace's method to the complex domain, the method of steepest descent, will be illustrated by applying it to approximate the Airy integral,

$$I(\sigma) = \int_{-\infty}^{\infty} e^{i\left(\sigma\xi - \frac{\xi^3}{3}\right)} d\xi , \tag{83}$$

where σ is assumed to be real.

First we study the convergence of this integral. Put $\xi^3 = t$; thus

$$I(\sigma) = \int_{-\infty}^{\infty} e^{i\left(\sigma t^{\frac{1}{3}} - \frac{t}{3}\right)} \frac{dt}{3t^{2/3}} . \tag{84}$$

The integral does not converge absolutely, since

$$\int_{-\infty}^{\infty} \frac{dt}{t^{2/3}} = \infty .$$

However, because of the imaginary exponential, the integrand in equation (84) is oscillatory, and because

$$\lim_{t \to \infty} \left(\frac{1}{t^{2/3}}\right) \to 0 ,$$

we can show that the integral converges. We integrate by parts between finite limits to obtain

$$\int_{-A}^{A} e^{-it/3} \frac{e^{i\sigma t^{\frac{1}{3}}}}{t^{2/3}} dt$$

$$= 3ie \frac{-it/3}{t^{2/3}} e^{i\sigma t^{\frac{1}{3}}} \Bigg|_{-A}^{A} - 3i \int_{-A}^{A} e^{-i(t/3)+i\sigma t^{\frac{1}{3}}} \left[\frac{-2}{3t^{5/3}} + \frac{i\sigma}{3t^{4/3}}\right] dt . \tag{85}$$

As $A \to \infty$ the first term on the right-hand side goes to zero and the remaining integral is absolutely convergent; this shows that the Airy integral (83) is convergent.

We attempt to evaluate the Airy integral by the method previously developed in Section 3.2. If we substitute

$$w = \sigma\xi - \frac{\xi^3}{3} , \qquad d\xi(\sigma - \xi^2) = dw ,$$

then (83) becomes

$$I = \int e^{iw} \frac{dw}{\sigma - \xi^2} . \tag{86}$$

From Figure 3.2, we see that ξ is not a single-valued function of w.

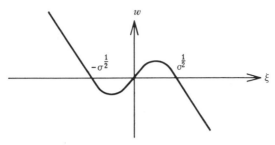

Figure 3.2

In order to make the graph dimensionless we use a scale transformation.

$$\xi = \sqrt{\sigma}\, z\, . \tag{87}$$

Substituting (87) into $I(\sigma)$ we have

$$I(\sigma) = \sqrt{\sigma} \int_{-\infty}^{\infty} e^{i\sigma^{\frac{3}{2}}[z - (z^3/3)]}\, dz\, . \tag{88}$$

We shall consider $I(\sigma)$ over the complex plane. Setting

$$w = z - \frac{z^3}{3} = u + iv\, , \tag{89}$$

the absolute value of the exponential is then

$$\left| e^{i\sigma^{\frac{3}{2}} w} \right| = e^{-\sigma^{\frac{3}{2}} v} \tag{90}$$

As $|w| \to \infty$ the real part of w produces very rapid oscillations, and this makes the integral difficult to evaluate. Hence we want to choose a contour that will facilitate the evaluation of the integral for large w. We choose a contour on which u is constant in order to minimize the effects of oscillations, and we also try to have the contour go through the point where the absolute value of the integrand is as large as possible. The curves along which u is constant are orthogonal to those along which v is constant, so that along a contour for which u is constant the function v varies fastest and so does the absolute value of the integrand.

In terms of $z = x + iy$, we have $w'(z) = u_x - iu_y = v_y + iv_x$ (by the Cauchy-Riemann conditions) so that u and v are both stationary at a point x_0, y_0 such that $w'(x_0, y_0) = 0$.

The point x_0, y_0 is neither a maximum or a minimum (since u and v satisfy Laplace's equation), but it is a saddle point; that is, the appearance of the surface $v(x, y)$ above a Cartesian x, y plane is saddle shaped around the point x_0, y_0. We now seek the contour corresponding to the steepest path (a path along which u is constant) through the saddle point.

Differentiating w we obtain

$$\frac{\partial w}{\partial z} = 0 = 1 - z^2 ; \tag{91}$$

therefore the saddle points are located at $z = \pm 1$, that is, $x, y = \pm 1, 0$. We have

$$u + i v = x + iy - \frac{1}{3} (x^3 + 3ix^2y - 3xy^2 - i y^3),$$

$$u = x - \frac{1}{3} (x^3 - 3xy^2), \tag{92}$$

$$v = y - \frac{1}{3} (3x^2y - y^3).$$

Thus at $z = 1$ ($x = 1$, $y = 0$), we obtain $u = 2/3$, $v = 0$; similarly at $z = -1$ ($x = -1, y = 0$) we obtain $u = -2/3$, $v = 0$.

In order that the contour u = constant pass through the saddle point at $z = 1$, we set the constant equal to $2/3$. We shall see that this specifies two curves, and we shall select the one along which $e^{-\sigma^{\frac{1}{2}} v}$ decreases (v increases) as fast as possible with increasing distance from the point; on such a path the absolute value of $I(\sigma)$ depends mainly on the region around the saddle point. The contour $u = 2/3$ has the equation

$$x - \frac{1}{3} (x^3 - 3xy^2) = \frac{2}{3} , \tag{93}$$

from which

$$y^2 = \frac{2 - 3x + x^3}{3x} = \frac{(x + 2)(x - 1)^2}{3x} .$$

Thus for $x \approx 1$,

$$y^2 \approx (x - 1)^2 , \tag{94}$$

and for large x,

$$y^2 \sim \frac{x^3}{3x} = \frac{x^2}{3} ,$$

or

$$y \sim \pm \frac{x}{\sqrt 3} . \tag{95}$$

From (94) and (95) we see that the curve $u = 2/3$ has branches which intersect the x axis at $\pm 45°$. Along one branch v goes down rapidly as the saddle point is approached, and along the other v goes up rapidly. See Figure 3.3.

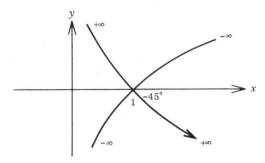

Figure 3.3

We have for large x,

$$v = y - \frac{1}{3} (3x^2 y - y^3) \approx y \left[1 - \frac{1}{3} \left(\frac{8x^2}{3} \right) \right],$$

and this goes to $\pm\infty$ as indicated in the above diagram. Thus for negative y, v approaches $+\infty$ with increasing x, so it is the branch at $-45°$ that we want.

At the other saddle point $z = -1$ ($x = -1$, $y = 0$) we have $u = -\frac{2}{3}$ and $v = 0$. The contour is now specified by

$$u = x - \frac{1}{3} (x^3 - 3xy^2) = -\frac{2}{3} \tag{96}$$

which corresponds to Figure 3.4.

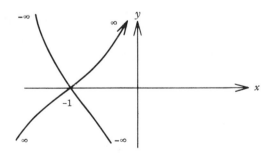

Figure 3.4

From Figures 3.3 and 3.4, we select the path of integration, shown bare in Figure 3.5. Along this path the variable part of w is imaginary, the variable part of iw is real, and the original integral separates into two real integrals.

For visualization, think of the terrain associated with the height of $e^{-\sigma^{\frac{3}{2}} v}$. The path of integration is from the valley at A through the high pass at B and down into the vally at C; then along the path from C through the high

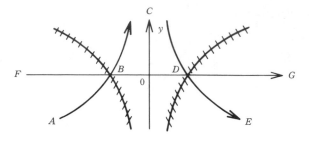

Figure 3.5

pass at D and down into the valley at E. The contour may be deformed to one going from $F(-\infty)$ through B through D to $G(\infty)$, the portions from 0 to C and C to 0 cancelling, to show that it is equivalent to the original path $-\infty$ to ∞ as in (83).

Returning to (88), we use $z - z_0 = z - (\mp 1) = s e^{\pm i\pi/4}$, $w_\pm = z_0 - z_0^3/3$ $= \pm 2/3$, $w''_\pm = -2z_0 = \mp 2$ in the neighborhoods of the stationary points and proceed essentially as for (73) - (82) to obtain

$$I = \sqrt{\sigma} \left[\int_{ABC} + \int_{CDE} \right] \exp \left[i\sigma^{3/2} w(z) \right] dz$$

$$= \sqrt{\sigma} \, \exp \left(-i \frac{2}{3} \sigma^{3/2} \right) \int_{-\infty}^{\infty} \exp \left[i\sigma^{3/2} 2 \left(s e^{i\pi/4} \right)^2 / 2 \right] e^{i\pi/4} \, ds$$

$$+ \sqrt{\sigma} \, \exp \left(i \frac{2}{3} \sigma^{3/2} \right) \int_{-\infty}^{\infty} \exp \left[-i\sigma^{3/2} 2 \left(s e^{-i\pi/4} \right)^2 / 2 \right] e^{-i\pi/4} \, ds$$

$$= \left[\exp \left(-i \frac{2}{3} \sigma^{3/2} + \frac{i\pi}{4} \right) + \exp \left(i \frac{2}{3} \sigma^{3/2} - \frac{i\pi}{4} \right) \right] \sqrt{\sigma} \int_{-\infty}^{\infty} \exp \left(-\sigma^{3/2} s^2 \right) ds$$

$$= 2 \sqrt{\frac{\pi}{\sigma^{1/2}}} \, \cos \left(\frac{2}{3} \sigma^{3/2} - \frac{\pi}{4} \right). \tag{97}$$

We derive this more directly in a subsequent section.

The following summary will serve to consolidate the key ideas. Given $I = \int e^{-Kf(z)} g(z) \, dz$, we use $f(z) = w$ and obtain

$$I = \int e^{-Kw} g(z) \, \frac{dw}{f'(z)} .$$

The points where $f'(z) = 0$ are branch points of w, and these must be watched in the w plane. The contour is deformed in the w plane in order to obtain the most rapid decrease in the absolute value of the integrand with increasing distance from these points. After a change of variable is made in the integral, Watson's lemma is applied to get the asymptotic expansion.

3.4 METHOD OF STATIONARY PHASE

The asymptotic evaluation of integrals containing a pure imaginary exponent can be handled by the method of stationary phase. The typical case is

$$I = \int_{-\infty}^{\infty} e^{iKf(x)} g(x) \, dx \tag{98}$$

where $f(x)$, $g(x)$ are real functions, and K is a large parameter.

When K is large the exponential function $e^{iKf(x)}$ will in general oscillate very rapidly; if g changes slowly, the rapid phase changes will cancel out values of the integrand, and $I \sim 0$. However, the value of the integral will become significant in regions where the exponential function $e^{iKf(x)}$ is oscillating very slowly. This occurs at points for which $f(x)$ is stationary, that is, where $f'(x) = 0$.

The first step is to subdivide the interval of integration so that we can examine the behavior of the integral when $f(x)$ has no stationary points. Hence consider

$$I_{ab} = \int_{a}^{b} e^{iKf(x)} g(x) \, dx \tag{99}$$

for $a \le x \le b$ and assume $f'(x) \ne 0$ in that interval. For simplicity assume $f'(x) > 0$ in the interval a to b, so that $f(x)$ is monotonic nondecreasing on (a, b) as in Figure 3.6.

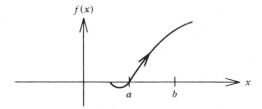

Figure 3.6

The integrand can be put into a more usable form by substituting $t = f(x)$, $dx = dt/f'(x)$. Then the integral becomes

$$I_{ab} = \int_{f(a)}^{f(b)} e^{iKt} \frac{g(x)}{f'(x)} \, dt \,. \tag{100}$$

From the diagram it is clear that the inverse of f exists so that $x = x(t)$ can be obtained.

Integrating by parts, we find

$$I_{ab} = g \, \frac{e^{iKt}}{iKf} \bigg|_{f(a)}^{f(b)} - \frac{1}{iK} \int_{f(a)}^{f(b)} e^{iKt} \frac{d}{dt} \left(\frac{g}{f'} \right) \, dt \,. \tag{101}$$

The first term has the order of magnitude $1/K$, and the integral on the right is also $O(1/K)$. This can be seen by differentiating g/f' to obtain $(f'g' - gf'')/(f')^2$; since, by assumption, $f' \neq 0$, the derivative is a bounded function of t. Hence the integral exists and is bounded and satisfies the same conditions as the original integral (99); the integral is thus $O(1/K^2)$. Thus whenever $f(x)$ is nonstationary and the parameter K is large, the sum of all these portions ΣI_{ab} on which $f'(x) \neq 0$ is $O(1/K)$.

We now proceed to show that the integral over an interval containing a stationary point is $O(1/\sqrt{K})$. Consider the integral for the vicinity of a stationary point x_0:

$$I_{x_0} = \int_{x_0-\epsilon}^{x_0+\epsilon} e^{iKf(x)} g(x)\, dx = \left[\int_{x_0-\epsilon}^{x_0} + \int_{x_0}^{x_0+\epsilon} \right] e^{iKf(x)} g(x)\, dx . \tag{102}$$

If $f'(x_0) = 0$ and $f''(x_0) \neq 0$, the curve $f(x)$ has a maximum or a minimum and can be approximated by a parabola in the neighborhood of x_0. As before for (75), we have

$$f(x) = f(x_0) + f'(x_0)(x - x_0) + \frac{f''(x_0)}{2!}(x - x_0)^2 + \cdots . \tag{103}$$

For the small region about x_0 we write

$$f(x) \approx f(x_0) + \frac{f''(x_0)}{2!}(x - x_0)^2 , \tag{104}$$

and change variables to τ ranging from 0 to η as determined by

$$f(x) = f(x_0) + \frac{1}{2} f''(x_0)\tau^2 , \quad \tau \approx x - x_0 . \tag{105}$$

The first derivative of $f(x)$ is then $f'(x)\, dx = f''(x_0)\tau\, d\tau$, so that the integral in (102) for the half-region x_0 to $x_0 + \epsilon$ becomes

$$I_{1/2} = e^{iKf(x_0)} \int_0^\eta \exp\left[iK \frac{f''(x_0)\tau^2}{2} \right] \frac{\tau f''(x_0)}{f'(x)} g(x)\, d\tau . \tag{106}$$

The function $\tau[f''(x_0)/f'(x)]g(x)$ can be expanded in a series about x_0 and then converted into a function of τ:

$$\frac{\tau f''(x_0)\, g(x)}{f'(x)} = h(\tau) . \tag{107}$$

We change scale by setting $\tau = \sigma/\sqrt{K}$, and obtain

$$I_{1/2} = e^{iKf_0} \int_0^\eta e^{iK \frac{f''_0 \tau^2}{2}} h(\tau)\, d\tau$$

$$= \frac{e^{iKf_0}}{\sqrt{K}} \int_0^{\sqrt{K}\eta} e^{i \frac{f''_0}{2}\sigma^2} h\left(\frac{\sigma}{\sqrt{K}} \right) d\sigma , \tag{108}$$

where $f_0 = f(x_0)$. The function $h(\sigma/\sqrt{K})$ tends to a constant $h(0)$ as K tends to infinity, so that the integrand becomes independent of K. Thus as K tends to infinity,

$$I_{\frac{1}{2}} \sim \frac{h(0)\,e^{iKf_0}}{K^{\frac{1}{2}}} \int_0^\infty \exp\!\left(i\frac{f''_0}{2}\sigma^2\right) d\sigma = \frac{h(0)\,e^{iKf_0}}{\sqrt{K}}\sqrt{\frac{\pi}{2\,|f''_0|}}\;e^{i(\pi/4)\,\mathrm{sgn}\,f''_0}\,. \tag{109}$$

where we used the well-known result for the Fresnel integral. Thus we see that $I_{\frac{1}{2}} = O(1/\sqrt{K})$, and hence $I_{\frac{1}{2}} > O(1/K)$.

In replacing the upper limit in (108) by ∞, we implicitly neglected the integral in the interval $\eta\sqrt{K}$ to ∞. The integral in this region is represented by

$$\Delta I_1 = \frac{e^{iKf_0}}{\sqrt{K}} \int_{\eta\sqrt{K}}^\infty \exp\!\left(if''_0\frac{\sigma^2}{2}\right) h\!\left(\frac{\sigma}{\sqrt{K}}\right) d\sigma\,, \tag{110}$$

and the previous argument shows that

$$\Delta I_1 = O\!\left(\frac{1}{K}\right) < O\!\left(\frac{1}{\sqrt{K}}\right)\,,$$

so that the result is indeed negligible compared to (109). In addition, on replacing $h(\sigma/\sqrt{K})$ by h_0, we neglected

$$\Delta I_2 = \frac{1}{\sqrt{K}}\int_0^\infty \exp\!\left(if''_0\frac{\sigma^2}{2}\right)\!\left[h\!\left(\frac{\sigma}{\sqrt{K}}\right) - h(0)\right] d\sigma = O\!\left(\frac{1}{K}\right); \tag{111}$$

the integral is $O(1/K)$ because by the mean value theorem

$$\left[h\!\left(\frac{\sigma}{\sqrt{K}}\right) - h(0)\right] = \frac{\sigma}{\sqrt{K}}\,h'\!\left(\frac{\theta\sigma}{\sqrt{K}}\right) = O\!\left(\frac{1}{\sqrt{K}}\right)\,,\quad 0 < \theta < 1. \tag{112}$$

Thus $I_{\frac{1}{2}}$ of (109) is the leading term in the expansion, and the development will be completed when $h(0)$ is determined. Since

$$f'(x) \approx f''_0\,\tau\,, \tag{113}$$

we have

$$h(0) = \lim_{\substack{x \to x_0 \\ \tau \to 0}} \frac{\tau f''_0\,g(x)}{\tau f''_0} = g(x_0)\,. \tag{114}$$

The integral for the region $x_0 - \epsilon$ to x_0 of (102) can be evaluated by integrating from minus infinity to zero to again obtain (109). Hence the total result is simply twice (109):

$$I_{x_0} = \frac{e^{iKf(x_0)}}{\sqrt{K}}\,g(x_0)\sqrt{\frac{2\pi}{|f''(x_0)|}}\;e^{i(\pi/4)\,\mathrm{sgn}\,f''(x_0)}\,. \tag{115}$$

Similarly, if there are several stationary points, the total effect is

$$I = \sum_{n=0}^{N} I_{x_n} = \sum_{n=0}^{N} \frac{e^{iKf(x_n)}}{\sqrt{K}} g(x_n) \sqrt{\frac{2\pi}{|f''(x_n)|}} \exp\left[i\frac{\pi}{4} \operatorname{sgn} f''(x_n)\right]. \quad (116)$$

Another method for evaluating the integral at stationary points x_0 is the following: first expand the functions $f(x)$ and $g(x)$ in their Taylor series about x_0 to obtain

$$I_{x_0} = \int_{-\infty}^{\infty} \exp\left\{iK\left[f(x_0) + \frac{f_0''}{2}(x - x_0)^2 + \cdots\right]\right\}$$

$$\times \left[g(x_0) + g'(x_0)(x - x_0) + \cdots\right] dx$$

$$= e^{iKf_0} g_0 \int_{-\infty}^{\infty} \exp\left\{i\frac{K}{2} f_0''(x - x_0)^2 + \cdots\right\}$$

$$\times \left[1 + \frac{g_0'}{g_0}(x - x_0) + \cdots\right] dx = I_0, \quad (117)$$

where we use $g_0 = g(x_0)$, etc., for brevity. Substituting $x - x_0 = \xi/\sqrt{K}$, we have

$$I_0 = e^{iKf_0} g_0 \int_{-\infty}^{\infty} \exp\left\{iK\left[\frac{f_0''}{2}\frac{\xi^2}{K} + \frac{f_0'''}{3!}\frac{\xi^3}{K^{3/2}} + \cdots\right]\right\}$$

$$\times \left[1 + \frac{g_0'}{g_0}\frac{\xi}{K^{1/2}} + \cdots\right] \frac{d\xi}{\sqrt{K}} \quad (118)$$

$$= \frac{e^{iKf_0}}{\sqrt{K}} \int_0^{\infty} e^{if_0''\xi^2/2} \left[1 + \frac{g_0'}{g_0}\frac{\xi}{\sqrt{K}} + \frac{if_0'''\xi^3}{3!\sqrt{K}} + O\left(\frac{1}{K}\right)\right] d\xi$$

where we brought down all but the f'' term from the exponent. The first term of (118) is $O(1/\sqrt{K})$, and all succeeding terms are $O(1/K)$ or smaller. If all but the $O(1/\sqrt{K})$ term are discarded, the final result is

$$I_0 \sim \frac{e^{iKf_0} g_0}{\sqrt{K}} \int_{-\infty}^{\infty} \exp\left(if_0''\xi^2/2\right) d\xi, \quad (119)$$

which reduces to (115) as before.

The above procedure breaks down at an inflection point where $f''(x_0) = 0$; here the appropriate change of scale is $x - x_0 = \xi/K^{1/3}$ and the corresponding integral of (117) becomes

$$I_0 \sim e^{iKf_0} g_0 \int_{-\infty}^{\infty} e^{iKf_0'''(x-x_0)^3/3!} dx$$

(120)

$$= \frac{e^{iKf_0} g_0}{K^{1/3}} 2 \, \mathrm{Re} \int_0^{\infty} e^{if_0''' \xi^3/3!} d\xi,$$

where, by Cauchy's theorem, we may replace the path 0 to ∞ by 0 to $\infty \, e^{i(\pi/6) \, \mathrm{sgn} \, f_0'''}$. Letting $\xi = e^{i(\pi/6) \, \mathrm{sgn} \, f_0'''} \eta^{1/3}$, we obtain

$$I_0 \sim \frac{e^{iKf_0} g_0}{K^{1/3}} 2 \, \mathrm{Re} \, e^{i(\pi/6) \, \mathrm{sgn} \, f_0'''} \int_0^{\infty} \frac{e^{-|f_0'''| \eta/6}}{3\eta^{2/3}} d\eta$$

(121)

$$= \frac{e^{iKf_0} g_0}{K^{1/3}} \frac{2 \cos (\pi/6)}{3} \left(\frac{6}{|f_0'''|} \right)^{1/3} \Gamma\left(\frac{1}{3}\right) = \frac{\Gamma\left(\frac{1}{3}\right)}{\sqrt{3}} \left(\frac{6}{K |f_0'''|} \right)^{1/3} g_0 \, e^{iKf_0},$$

where the value of the gamma function $\Gamma(z)$ for $z = 1/3$ is approximately $\Gamma(1/3) \approx 2.68$.

The following examples will serve to illustrate the method of stationary phase. First we will show how to approximate the Airy integral (83) and again obtain the result (97). We start with

$$I = \int_{-\infty}^{\infty} \exp\left[i\left(\sigma x - x^3/3\right)\right] dx, \quad \sigma > 0,$$

(122)

and change variables to $x = \sqrt{\sigma} y$, where y is also real:

$$I = \sqrt{\sigma} \int_{-\infty}^{\infty} \exp\left[i\sigma^{3/2}\left(y - \frac{y^3}{3}\right)\right] dy.$$

(123)

Here the function $g(x)$ has the value unity. We set $K = \sigma^{3/2}$ to get the standard form as in (98):

$$I = K^{1/3} \int_{-\infty}^{\infty} \exp\left[iK\left(y - \frac{y^3}{3}\right)\right] dy = K^{1/3} \int_{-\infty}^{\infty} e^{iKf(y)} dy$$

(124)

The stationary points y_n are determined by setting the derivative of the exponent equal to zero. We have $f_n'(y) = 1 - y_n^2 = 0$, and consequently $y_n = \pm 1$. The corresponding values of the second derivative are $f''(y_n) = 2y_n \neq 0$, and the appropriate values of the phase are determined by $f(y_n) = \pm(1 - 1/3) = \pm 2/3$.

Equation (116) can now be applied to obtain

$$I \sim K^{1/3} e^{2iK/3} \sqrt{\frac{2\pi}{|-2|K}} \; e^{-i\pi/4} + K^{1/3} e^{-2iK/3} \sqrt{\frac{2\pi}{2K}} \; e^{i\pi/4} , \quad (125)$$

or equivalently,

$$I \sim \sigma^{1/2} \sqrt{\frac{\pi}{\sigma^{3/2}}} \; 2 \cos \left(\frac{2}{3} \sigma^{3/2} - \frac{\pi}{4} \right) \qquad (126)$$

as in (97).

As a second example we consider the zeroth-order Bessel function. We have

$$J_0 (r) = \frac{1}{\pi} \int_0^\pi \exp (ir \cos \theta) \, d\theta . \qquad (127)$$

Here $f (\theta) = \cos \theta$, $f' (\theta) = - \sin \theta$, and $f'' (\theta) = - \cos \theta$. The points for which $f' (\theta) = 0$ are $\theta = 0$ and $\theta = \pi$, and since these stationary points are located at the end points, their contributions to equation (116) are halved. Thus the values of the integral at the stationary points are

$$\int_{\theta=0} \sim \frac{1}{2} \frac{1}{\pi} e^{ir} \sqrt{\frac{2\pi}{r}} \; e^{-i\pi/4}$$

$$\int_{\theta=\pi} \sim \frac{1}{2} \frac{1}{\pi} e^{-ir} \sqrt{\frac{2\pi}{r}} \; e^{i\pi/4}, \qquad (128)$$

and the sum of the two is the required result:

$$J_0 (r) \sim \sqrt{\frac{2}{\pi r}} \left\{ \frac{\exp \left[i \left(r - \frac{\pi}{4} \right) \right] + \exp \left[-i \left(r - \frac{\pi}{4} \right) \right]}{2} \right\}$$

$$= \sqrt{\frac{2}{\pi r}} \cos \left(r - \frac{\pi}{4} \right) . \qquad (129)$$

Consider another integral of the Airy type

$$I = \int_{-\infty}^\infty \exp \left[iK \left(ay - \frac{y^3}{3} \right) \right] \, dy . \qquad (130)$$

The stationary points corresponding to $f' (y) = a - y^2 = 0$ are $y = \pm \sqrt{a}$, and the second derivative has the values $f'' = -2y = \mp 2\sqrt{a}$. When $a \to 0$ the stationary points coalesce to one point $y = 0$. In this case $f'(y) = f''(y) = 0$ at the stationary point, and we expand up to cubic terms and use equation (120) instead of (116). As an application in wave physics, the form (116) for $a \neq 0$ suffices for individual geometrical rays of a set except on the envelope of the set of rays. The envelope (the caustic) corresponds to $f'' = \mp 2\sqrt{a} \to 0$, and the result of (116) would then become singular; however, this merely indicates that (116) is inappropriate for $f'' = 0$ and that we require (121).

TWO DIMENSIONS

The method of stationary phase will now be extended to two-dimensional integrals of the form

$$I = \iint e^{iKf(x,y)} g(x,y) dx\, dy \tag{131}$$

in order to obtain approximations for large K. We set $f(x,y) = p$, and we regard the level lines as given by the equations corresponding to constant values of p. The trajectories orthogonal to the curves along which $f(x,y)$ is constant will be designated by $\phi(x,y) = q$ for constant values of q. The level lines and the orthogonal trajectories will be used as a new coordinate system (p,q). Since

$$dx\, dy = J\, dp\, dq , \tag{132}$$

$$J = \frac{\partial(x,y)}{\partial(p,q)} = \frac{1}{\begin{vmatrix} f_x & f_y \\ \phi_x & \phi_y \end{vmatrix}} = \frac{1}{f_x\,\phi_y - f_y\,\phi_x}$$

where J is the Jacobian, we rewrite (131) in the form

$$I = \iint e^{iKp} gJ\, dp\, dq . \tag{133}$$

The slope of a level line $f(x,y) = p$ is obtained from $f_x + f_y\,(dy/dx) = f_x + f_y\, y' = 0$,

$$y' = -\frac{f_x}{f_y} , \tag{134}$$

and the slope of an orthogonal trajectory $\phi(x,y) = q$ through the same point (that is, the curve perpendicular to the level line) equals $y' = f_y/f_x$. We also have $\phi_x + \phi_y y' = 0$, from which

$$y' = -\frac{\phi_x}{\phi_y} . \tag{135}$$

Thus for the orthogonal trajectories,

$$y' = \frac{f_y}{f_x} = -\frac{\phi_x}{\phi_y} . \tag{136}$$

Substituting this into J^{-1}, we obtain

$$J^{-1} = f_x\,\phi_y - f_y\,\phi_x = \frac{\phi_y}{f_x}\left[f_x^{\,2} + f_y^{\,2}\right]. \tag{137}$$

and the integral (133) becomes

$$I = \int\int e^{iKp} \frac{g(x,y) f_x}{\phi_y} \frac{dp \, dq}{\left[f_x^2 + f_y^2\right]} . \tag{138}$$

If the function $f(x,y)$ is well behaved, so that the integrand has no singularities, then the analog of the procedure for the one-dimensional case (101) now gives $I = O(1/K^2)$. However, if there are stationary points at which $f_x = f_y = 0$, then the corresponding analog of the one-dimensional result (116) is now $I = O(1/K)$. For the stationary case near a point x_0, y_0 at which

$$f_y(x_0,y_0) = 0, \qquad f_x(x_0,y_0) = 0, \tag{139}$$

we expand the integrand of (131) in a Taylor series to obtain the analog of (117) and keep only the leading term of g and up to quadratic terms in f:

$$I = e^{iKf(x_0,y_0)} g(x_0,y_0)$$

$$\tag{140}$$

$$\times \int\int \exp\left\{i \frac{K}{2}\left[f_{xx}(x-x_0)^2 + 2f_{xy}(x-x_0)(y-y_0) + f_{yy}(y-y_0)^2\right]\right\} dx \, dy.$$

We now let $x - x_0 = \xi/\sqrt{K}$, $y - y_0 = \eta/\sqrt{K}$, and the integral takes a form similar to the one-dimensional case (119):

$$I = e^{iKf_0} g_0 \frac{1}{K} \int\int \exp\left[\frac{1}{2}\left(f_{xx}\xi^2 + 2f_{xy}\xi\eta + f_{yy}\eta^2\right)\right] d\xi \, d\eta. \tag{141}$$

The cross term in the exponent can be eliminated by a suitable coordinate rotation, and I is then a product of integrals of the Fresnel type.

FINITE LIMITS

We again consider the integral

$$I = \int_a^b e^{iKf(x)} g(x) \, dx, \tag{142}$$

where the functions $f(x)$ and $g(x)$ are real and K is a large parameter. If the region of integration (a,b) contains a stationary point $a < x_0 < b$ at which $f'(x_0) = 0$ and $f''(x_0) \neq 0$, than, as discussed for (115),

$$I \sim e^{iKf(x_0)} g(x_0) \sqrt{\frac{2\pi}{|Kf''(x_0)|}} \exp\left[i\frac{\pi}{4} \text{sgn} Kf''(x_0)\right] \tag{143}$$

plus terms of $O(1/K)$. Equation (143) would be summed over the stationary points, if there were more than one.

We may parallel the previous considerations of (117) and (118) to obtain an approximation of (142) which preserves the finite limits. Thus in terms of $\xi = \sqrt{K}\,(x-x_0)$, we obtain

$$I \sim \frac{e^{iKf_0}g_0}{\sqrt{K}} \int_{\sqrt{K}\,(a-x_0)}^{\sqrt{K}\,(b-x_0)} \exp\left(if_0''\,\xi^2/2\right)d\xi, \tag{144}$$

where, as K tends to infinity, one limit goes to $+\infty$ and the other goes to $-\infty$, since we have assumed $a < x_0 < b$.

Equation (144) expresses the result in terms of the tabulated Fresnel integral. For convenience, and to show the relation with (119), we change variables to $y^2 = |f_0''|\,\xi^2/2$ and obtain

$$I = e^{iKf(x_0)}\,g(x)\left(\frac{2}{|Kf''(x_0)|}\right)^{\frac{1}{2}} \int_{-\alpha}^{\beta} \exp(iy^2\,\mathrm{sgn}f_0'')dy, \tag{145}$$

where

$$\alpha = \left(\frac{|Kf''(x_0)|}{2}\right)^{\frac{1}{2}}(a - x_0),$$

$$\beta = \left(\frac{|Kf''(x_0)|}{2}\right)^{\frac{1}{2}}(b - x_0),$$

which is also valid if x_0 equals either a or b.

3.5 AIRY INTEGRAL

We now return to the Airy integral and consider it in greater detail. We work with

$$A(K) = \int_{-\infty}^{\infty} e^{iK(3\alpha x - x^3)}dx \tag{146}$$

for large K. Following the routine procedure we used for (124) etc., we have for $\alpha > 0$,

$$f(x) = (3\alpha x - x^3);$$

$$f'(x) = 3(\alpha - x^2), \qquad x_0 = \pm\alpha^{1/2}; \tag{147}$$

$$f''(x) = -6x.$$

Thus from (143), using $x_0 = \alpha^{1/2}$ and $x_0 = -\alpha^{1/2}$, we get

$$A(K) ~\sim~ e^{iK2\alpha^{3/2}} \left(\frac{2\pi}{6\alpha^{1/2}K} \right)^{1/2} e^{-i\pi/4}$$

$$+ ~ e^{iK2\alpha^{3/2}} \left(\frac{2\pi}{6\alpha^{1/2}K} \right)^{1/2} e^{i\pi/4}, \tag{148}$$

or equivalently

$$A(K) ~\sim~ \left(\frac{2\pi}{6\alpha^{1/2}K} \right)^{1/2} 2 \cos \left(2K\alpha^{3/2} - \pi/4 \right) \tag{149}$$

which correspond to (125) and (126), respectively.

This method breaks down when the two stationary points coalesce, that is, if $\alpha = 0$; then the second derivative of $f(x)$ is zero at the stationary point $x_0 = 0$, and we must work with a more complete expansion in the neighborhood of $x_0 = 0$. Corresponding to (146) for $\alpha = 0$, we use

$$f(x) = -x^3 = f(x_0) + f_0''' \frac{(x - x_0)^3}{3!}, \tag{150}$$

Proceeding as for (120), we set $x = yK^{-1/3}$, to obtain

$$A(K) = \int_{-\infty}^{\infty} e^{-iKx^3} dx = \frac{1}{K^{1/3}} 2\mathrm{Re} \int_{0}^{\infty} e^{-iy^3} dy, \tag{151}$$

which is just a special case of (120). As before, we set $y = e^{-i\pi/6} u^{1/3}$, so that

$$A = \frac{2}{K^{1/3}} \frac{\mathrm{Re}\, e^{-i\pi/6}}{3} \int_{0}^{\infty} e^{-u} u^{-2/3} du = \frac{2}{3K^{1/3}} \mathrm{Re}\, e^{-2\pi/6} \Gamma\left(\frac{1}{3}\right)$$

$$= \frac{\Gamma\left(\frac{1}{3}\right)}{\sqrt{3}\, K^{1/3}}, \tag{152}$$

which follows directly from (121) by setting $g_0 = 1$, $f_0 = 0$, and $|f_0'''|/6 = 1$.

The coalescence of stationary points occurs physically on caustics in optics, and on the cusp line for waves in the wake of a ship. On such lines, the stationary phase points (individually of order $1/\sqrt{K}$) that have coalesced produce a larger amplitude of order $1/K^{1/3}$.

The method of expanding the function $f(x)$ in a power series and neglecting higher-order terms does not always work. For example, consider the integral

$$B = \int_{-\infty}^{\infty} e^{i(3x^2 - 2\epsilon x^3)} \, dx \, . \tag{153}$$

We shall show that the limit for $\epsilon \to 0$ is not asymptotically equal to

$$\int_{-\infty}^{\infty} e^{i3x^2} \, dx = \sqrt{\frac{\pi}{3}} \, e^{i\pi/4} \, . \tag{154}$$

We change the scale by setting $x = z/\epsilon$ and $K = \epsilon^{-2}$ to obtain

$$B = \int_{-\infty}^{\infty} e^{iK(3z^2 - 2z^3)} K^{1/2} \, dz \, . \tag{155}$$

Applying the method of stationary phase, we have

$$f(z) = 3z^2 - 2z^3 \, ,$$

$$f'(z) = 6(z - z^2) \, ,$$

$$f''(z) = 6 - 12z \, .$$

The points of stationary phase occur at $z = 0$ and $z = 1$. Thus using (116) we obtain

$$B \sim K^{1/2} \left[\sqrt{\frac{2\pi}{6K}} \, e^{i\pi/4} + e^{iK} \sqrt{\frac{2\pi}{6K}} \, e^{-i\pi/4} \right] = \sqrt{\frac{\pi}{3}} \left[e^{i\pi/4} + e^{i(K - \pi/4)} \right], \tag{156}$$

or equivalently,

$$B \sim \sqrt{\frac{\pi}{3}} \, 2e^{iK/2} \cos\left(\frac{K}{2} - \frac{\pi}{4} \right), \tag{157}$$

an expression that has no limit as $K - 1/\epsilon^2 \to \infty$. Equation (156) consists of (154) plus a term arising from the stationary point $z = \epsilon x = 1$; the stationary point at $z = 1$ was neglected in (154) by setting $\epsilon = 0$ at the start. As a general rule we drop terms in the exponent containing a small parameter only if we do not eliminate some stationary point in the process.

CONTOUR INTEGRATION PROCEDURES

We return to the consideration of the Airy integral for the case $\alpha < 0$. Before proceeding we shall discuss an example that serves to review contour integration around a branch point singularity.

Consider the integral

$$J = \int_{\alpha}^{\beta} e^{iKx^2} \, dx. \tag{158}$$

By setting $x^2 = t$ we have

$$J = \int_{\alpha^2}^{\beta^2} e^{iKt} \frac{dt}{2t^{1/2}} ,$$ (159)

which has a branch point singularity at $t = 0$. Initially, we consider the integral to be taken along some contour from α^2 to β^2 that does not lie below $t = 0$; see Figure 3.7.

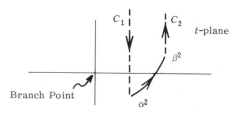

Figure 3.7

To obtain an asymptotic expansion of (159) by Watson's lemma, it is useful to have an integrand with a decaying exponential. This will occur if we evaluate the integral over the lines C_1 and C_2 which are parallel to the imaginary axis. By Cauchy's theorem we find that the integral around the entire contour C_1, α^2 to β^2, C_2 is zero, that is,

$$\int_C = \int_{C_1} + \int_{\alpha^2}^{\beta^2} + \int_{C_2} = 0 ,$$ (160)

because no singularities are enclosed by the contour. Thus the required integral from α^2 to β^2 may be written as

$$J = \int_{\alpha^2}^{\beta^2} = -\int_{C_1} - \int_{C_2} .$$ (161)

If we change the variable of integration on C_1 to $t = \alpha^2 + i\tau$, the integral along C_1 becomes

$$\int_{C_1} = -e^{iKa^z} \int_0^\infty e^{-K\tau} \frac{i\,d\tau}{2\sqrt{\alpha^2 + i\tau}}$$ (162)

and this can be evaluated asymptotically by using Watson's lemma as in (49). A similar transformation $t = \beta^2 + i\tau$ can be used on C_2. After the integrals $\int_{C_1}$ and $\int_{C_2}$ are evaluated, the desired integral from α^2 to β^2 as in (159) can be obtained from (161).

Now suppose the contour from α^2 to β^2 passes under the branch point $t = 0$, as shown in Figure 3.8.

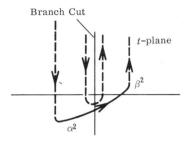

Figure 3.8

We again use Cauchy's theorem; since the contour integral encloses the branch point, we now have

$$\int_{\text{cut}} = \int_{C_1} + \int_{\alpha^2}^{\beta^2} + \int_{C_2} .$$ (163)

The integral along the branch cut is

$$\int_{\text{cut}} = \left[e^{i\pi} \int_{i\infty}^0 + \int_0^{i\infty} \right] e^{ikt} \frac{dt}{2t^{1/2}} = 2e^{i\pi/4} \int_0^\infty \frac{e^{-K\tau}}{2\sqrt{\tau}} d\tau = \sqrt{\frac{\pi}{K}} e^{i\pi/4}$$ (164)

where the factor $e^{i\pi}$ takes account of the phase change $(e^{i2\pi})^{1/2}$ of $\sqrt{t}$ on circling the branch point. The integrals along C_1 and C_2 can be evaluated as mentioned for (162). Thus when the contour passes under the branch point the solution consists of the two terms of (161) plus the integral (164) around the cut:

$$\int_{\alpha^2}^{\beta^2} = -\int_{C_2} - \int_{C_1} + \int_{\text{cut}} .$$

In the event that the contour α^2 to β^2 is along the real axis, as shown in Figure 3.9, it is not clear whether to integrate above or below the branch point.

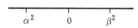

Figure 3.9

Suppose that $\beta^2 = 1$ and $\alpha^2 = -1$; then we might have $\beta = 1$, and either $\alpha = e^{i\pi/2}$ or $\alpha = e^{-i\pi/2}$. We shall see that these different values of α correspond to different ways of going around the branch point.

Since $t = x^2$, the value $t = \beta^2 = 1$ corresponds to $x = \pm\sqrt{1}$. We assume $x = +\sqrt{1}$, which just amounts to picking a certain branch of the square root.

In polar form, we have $t = \rho e^{i\theta}$, and $t^{1/2} = \sqrt{\rho} \, e^{i\theta/2}$. For $t = 1$, we require $\theta = 0$ since this gives the value $\beta = 1$. At $\alpha^2 = -1$, the angle θ has the values $\pm\pi$, and in going from α^2 to β^2 the phase of

$\sqrt{t}$ changes from $\theta/2 = \pm\pi/2$ to $\theta/2 = 0$, where the sign depends on which of the two possible contours we follow. If $\alpha = e^{i\pi/2}$, then in the t-plane θ goes from π to 0, and the integral should be taken along the contour C_3 of Figure 3.10. If $\alpha = e^{-i\pi/2}$, then θ goes from $-\pi$ to 0, and the integral should be taken along C_4. In either case we complete the contour by adding C_1 and C_2. Hence in the event the contour $C_1 + C_3 + C_2$ is chosen, no singular point is enclosed, and Cauchy's integral theorem gives the value zero: the required integral is obtained from equation (161). If the contour $C_1 + C_4 + C_2$ is chosen, the singular point is enclosed: the required integral is obtained from (165).

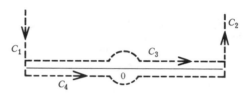

Figure 3.10

Now we return to the problem of evaluating the Airy integral (146) for $\alpha < 0$:

$$A = \int_{-\infty}^{\infty} e^{iK(3\alpha x - x^3)} dx, \qquad \alpha = -|\alpha| \tag{166}$$

Writing $t = 3\alpha x - x^3$, we see that $t'(x) = 3(\alpha - x^2) = -3(|\alpha| + x^2)$ does not vanish for real x; thus there are no real stationary points if $\alpha < 0$. In terms of t, the integral A takes the form

$$A = \int_{-\infty}^{\infty} e^{iKt}\left(\frac{dx}{dt}\right) dt \tag{167}$$

where the path is simply from $-\infty$ to ∞ because the transformation from x to t is monotonic, i.e., since $\alpha = -|\alpha|$, the slope dt/dx does not change sign.

In order to evaluate A by Watson's lemma we again try to choose a contour along which the integrand contains a decaying exponential. As before, such a contour parallel to the imaginary axis can be obtained by deforming the original contour upward from Re t. Since

$$\frac{dx}{dt} = \frac{1}{3(\alpha - x^2)} \tag{168}$$

has singular points at $x = \pm i|\alpha|^{1/2}$, dx/dt is singular in the upper half-plane; we cannot swing the contour up indiscriminately, but must avoid the branch point, as shown in Figure 3.11.

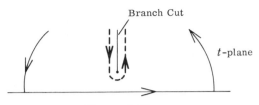

Figure 3.11

At the singular point $x_0 = -i|\alpha|^{1/2}$, where

$$t(x_0) = -(3|\alpha| + x_0^2)x_0 = -2|\alpha|(-i|\alpha|^{1/2}) = 2i|\alpha|^{3/2}, \qquad (169)$$

we make the change of variable $t = 2i\alpha^{3/2} + i\tau$. The Airy integral along the branch cut then takes the form

$$A = 2e^{-K2|\alpha|^{3/2}} \int_0^\infty e^{-K\tau} \frac{id\tau}{3(\alpha - x^2)}. \qquad (170)$$

The integrand $(\alpha - x^2)^{-1}$ can be expanded as a power series in τ and we can integrate termwise by Watson's lemma. The leading term is obtained directly from (166) by expanding the exponent to quadratic terms around the stationary point $x_0 = -i|\alpha|^{1/2}$; since $t_0 = i2|\alpha|^{3/2}$ and $t_0'' = -6x_0 = i6|\alpha|^{1/2}$, we have $iKt \approx iKt_0 + iKt_0''(x - x_0)^2/2 = -2K|\alpha|^{3/2} - 3K|\alpha|^{1/2}(x - x_0)^2$, and

$$A \sim e^{-2K|\alpha|^{3/2}} 2 \int_0^\infty e^{-3K|\alpha|^{1/2}s^2} ds = e^{-2K|\alpha|^{3/2}} \sqrt{\frac{\pi}{3K|\alpha|^{1/2}}} \qquad (171)$$

which decreases exponentially as $\alpha \to -\infty$. In Figure 3.12 we give a schematic representation of the behavior of the Airy integral for all values of α.

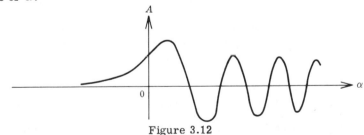

Figure 3.12

The Airy integral for $\alpha > 0$ has been considered previously by other methods. We shall now indicate how the same result could be obtained by the method of this section. We again make the substitution $t = 3\alpha x - x^3$ and try to find a suitable contour. As before, t is real when x is real, but now because the slope dt/dx is no longer constant in sign, the interval $-\infty < x < \infty$ does not go over simply into $-\infty < t < \infty$. From

$$\frac{dt}{dx} = 3(\alpha - x^2), \qquad (172)$$

we see that for $x < -\sqrt{\alpha}$, $dt/dx < 0$; for $-\sqrt{\alpha} < x < \sqrt{\alpha}$, $dt/dx > 0$; and for $x > \sqrt{\alpha}$, $dt/dx < 0$. In the t-plane the integrand of the Airy integral will have two branch points; $t = \pm 2\alpha^{3/2}$ corresponding to $x = \pm\sqrt{\alpha}$. For the integration, the value of t starts at $+\infty$ (for $x = -\infty$), decreases until $x = -\sqrt{\alpha}$ (where $t' = 0$), then increases until $x = +\sqrt{\alpha}$ (where again $t' = 0$), and then decreases to $-\infty$ (for $x = +\infty$). Hence, if we displace it slightly from the real axis, the contour in the t-plane appears as the dotted line shown below in Figure 3.13.

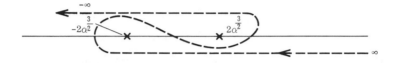

Figure 3.13

As always, we are seeking a contour along which the integrand contains a decreasing exponential function. Therefore we rotate the portion of the contour $t = -\infty$ to $t = 2\alpha^{3/2}$ until it becomes parallel to the imaginary axis. At the same time we push the part of the contour between $-2\alpha^{3/2}$ and $2\alpha^{3/2}$ in a direction parallel to the imaginary axis until it reaches infinity. The result is shown in Figure 3.14.

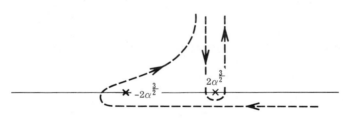

Figure 3.14

Since t was a cubic in x, we are working on a three-sheet Riemann surface, and the bottom portion of the contour is not on the same sheet as the branch point at $2\alpha^{3/2}$. Therefore this contour can also be moved until it is parallel to the imaginary axis, and the Airy integral reduces to the two integrals around the branch cuts of Figure 3.15.

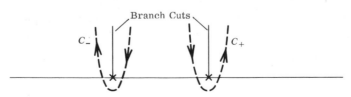

Figure 3.15

To illustrate the last argument explicitly, let $\alpha = 1$, so that $t = 3x - x^3$. Then the branch points are determined from

$$\frac{dt}{dx} = 3(1 - x^2) = 0, \qquad x = \pm 1 . \tag{173}$$

We see that at $x = 1$ we have $t = 2$, and at $x = -1$ we have $t = -2$. However, when a value of t is given, there are three values of x. Thus when a value of 2 is assigned to t, the three values of x are determined by solving the cubic

$$2 = 3x - x^3, \tag{174}$$

$$x^3 - 3x + 2 = 0 .$$

The root $x = 1$ is already known, so the remaining two roots can be determined by solving a quadratic in x. The solution is

$$(x - 1)^2 (x + 2) = 0, \tag{175}$$

$$x = 1, 1, -2 .$$

Tracing the values of x which correspond to the values of t on the original contour of integration, we find $x = -2$ on the lower curve. Since $x = -2$ is not a branch point, the lower curve can be pushed up. Thus starting from the contours of Figure 3.16,

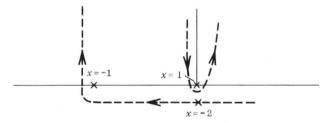

Figure 3.16

we may rotate the bottom curve to its position in Figure 3.17.

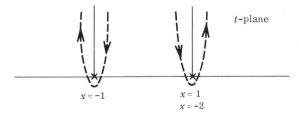

Figure 3.17

and thereby isolate the remaining branch cut as we did before for the more general case.

Thus we can write (166) and (167) for $\alpha > 0$ as the sum of two integrals around the branch cuts, i.e.,

$$A = \left[\int_{C-} + \int_{C+} \right] \frac{e^{iKt}}{3(\alpha - x^2)} \, dt \; ;$$

and substitute $t = \pm 2\alpha^{3/2} + i\tau$ in the corresponding integrands of $C_\pm$ to obtain versions for expanding by Watson's lemma. The leading terms are given in (148).

POLE NEAR STATIONARY POINT

Suppose a pole exists near a saddle point or a stationary phase point. For example, consider

$$D = \int_C \frac{e^{iKt}}{t^{1/2}} \frac{1}{t + \alpha} \, G(t) \, dt \qquad (176)$$

and the contour illustrated in Figure 3.18.

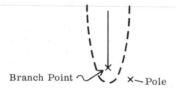

Branch Point ⟶ ✗ ✗⟵Pole

Figure 3.18

Changing the variable t to $i\tau$, we obtain

$$D = c \int_0^\infty \frac{e^{-K\tau}}{\tau^{1/2}} \frac{1}{(\tau - i\alpha)} \, G(i\tau) d\tau \; , \qquad c = 2e^{-i\pi/4} \qquad (177)$$

which is approximately

$$\frac{2G(0)e^{i\pi/4}}{\alpha} \int_0^\infty \frac{e^{-K\tau}}{\tau^{1/2}} \, d\tau = 2 \frac{G(0)}{\alpha} \frac{\pi}{K} \, e^{i\pi/4}$$

Further terms in an expansion of D may be obtained by applying Watson's lemma.

In the event that K is not large, or if α is small, our preceding approximation is not very good. In particular, if α is small the pole is very close to the branch point and contributes significantly to the integral; we try to isolate the difficulty as follows.

Suppose the function $G(t)$ is expressed as

$$G(t) = G(-\alpha) + \text{remainder}. \qquad (178)$$

The remainder can be evaluated by writing $G(t) - G(-\alpha) = (t + \alpha) G_1(t)$. This eliminates the pole at $t = -\alpha$. However, we must also cancel the singularity at $t = 0$ arising from the factor $1/t^{1/2}$. To take care of these two singularities we subtract from $G(t)$ a linear function which has the same values at $t = 0$ and $t = -\alpha$ as $G(t)$ has. Such a linear function is

$$G(-\alpha) + \frac{G(0) - G(-\alpha)}{\alpha} (t + \alpha) . \tag{179}$$

We write

$$G(t) - \left[G(-\alpha) + \frac{G(0) - G(-\alpha)}{\alpha} (t + \alpha) \right] = t(t + \alpha) G_2(t). \tag{180}$$

Then the integral D is approximately

$$G(-\alpha) \int \frac{e^{iKt}}{t^{1/2}} \frac{dt}{t + \alpha} + \frac{G(0) - G(-\alpha)}{\alpha} \int \frac{e^{iKt}}{t^{1/2}} dt , \tag{181}$$

and if we include the term in G_2, we have exactly

$$D = G(-\alpha) \int \frac{e^{iKt}}{t^{1/2}} \frac{dt}{t + \alpha} + \frac{G(0) - G(-\alpha)}{\alpha} \int \frac{e^{iKt}}{t^{1/2}} dt + \int e^{iKt} t^{1/2} G_2(t) \, dt. \tag{182}$$

The second integral on the right is of the Fresnel type, and the last integral is evaluated by use of Watson's lemma. Thus only the first seems to be a new form.

Consider the first integral

$$D_1 = \int \frac{e^{iKt}}{t^{1/2}} \frac{dt}{t + \alpha} . \tag{183}$$

Let $t = i\tau$ for $\alpha < 0$. Then the integral may be written

$$D_1 = c \int_0^\infty \frac{e^{-K\tau} d\tau}{\tau^{1/2}(\tau - i\alpha)} = -ic \int_0^\infty \frac{e^{-K\tau}}{\tau^{1/2}} d\tau \int_0^\infty e^{i(\tau - i\alpha)\sigma} \, d\sigma. \tag{184}$$

By interchanging the order of integration we obtain

$$D_1 = -ic \int_0^\infty d\sigma \, e^{\alpha\sigma} \int_0^\infty \frac{e^{-\tau(K - i\sigma)}}{\tau^{1/2}} d\tau , \tag{185}$$

which we reduce by integrating over the variable τ:

$$D_1 = -ic \int_0^\infty \frac{\pi^{1/2} e^{\alpha\sigma}}{(K - i\sigma)^{1/2}} d\sigma . \tag{186}$$

Substituting $u = K - i\sigma$, we obtain

$$D_1 = \pi^{1/2} e^{-i\alpha K} c \int_K^{-i\infty} \frac{e^{i\alpha u}}{u^{1/2}} du \tag{187}$$

which we recognize as a Fresnel integral that can be expanded by the methods discussed earlier.

CHAPTER 4

Difference Equations

4.1 RECURRENCE RELATIONS AND DIFFERENCE EQUATIONS

INTRODUCTION

Problems involving difference equations or recurrence relations arise naturally in applications where the variable assumes only a discrete sequence of values. We shall see that difference equations and recurrence relations are just two aspects of the same problem and that the methods of solution are very similar to those used in solving differential equations. Just as with differential equations, most difference equations and recurrence relations cannot be solved explicitly. However, we shall concentrate on those that can, and shall present general methods for treating them. Most of the exposition will be by means of examples, but it is hoped that the underlying ideas will be manifest.

Suppose we wish to evaluate the integral

$$I_n = \int_0^{\pi/2} \sin^n x \, dx, \tag{1}$$

where n is a positive integer. An integration by parts shows that

$$I_n = \int_0^{\pi/2} \sin^{n-1} x \sin x \, dx$$

$$= -\sin^{n-1} x \cos x \Big|_0^{\pi/2} + (n-1) \int_0^{\pi/2} \sin^{n-2} x \cos^2 x \, dx$$

$$= (n-1) \int_0^{\pi/2} \sin^{n-2} x \, (1 - \sin^2 x) \, dx$$

$$= (n-1) [I_{n-2} - I_n].$$

Simplifying, we get the recurrence relation

$$n I_n = (n-1) I_{n-2} , \tag{2}$$

or

$$I_n = \frac{n-1}{n} I_{n-2} = \frac{(n-1)(n-3)}{n(n-2)} I_{n-4} = \cdots \tag{3}$$

$$= \frac{(n-1)(n-3) \cdots}{n(n-2) \cdots} [I_1 \text{ or } I_0] ,$$

according as n is an odd or even integer. Since, obviously, $I_1 = 1$

110

and $I_0 = \pi/2$, formula (3) gives the explicit solution of (2) and the value of the integral (1).

As another example of a recurrence relation, consider the well-known problem of the gambler's ruin. Suppose the gambler makes a series of even bets (that is, probability of $1/2$ of winning) of one dollar each. He starts with a dollars and his opponent with b dollars. What is the probability of the gambler bankrupting his opponent?

Let $f(n)$ denote the probability of the gambler bankrupting his opponent, when the gambler has n dollars and the opponent has $a + b - n$ dollars; the probability we want is $f(a)$, where a is the amount he started with. We may easily obtain a recurrence relation for $f(n)$ as follows:

When the gambler has n dollars, and the probability of bankrupting his opponent is $f(n)$, the sequence may continue in either of two ways: with probability $1/2$ he loses the next bet, thus leaving himself with $n - 1$ dollars, and then, with probability $f(n - 1)$, he bankrupts his opponent; or, with probability $1/2$, he wins the next bet, thus obtaining $n + 1$ dollars, and then, with probability $f(n + 1)$, he bankrupts his opponent. Combining these facts, we see that

$$f(n) = \frac{1}{2} f(n - 1) + \frac{1}{2} f(n + 1) ; \qquad (4)$$

This is the desired recurrence relation. However, to give a complete description of our particular problem, we must add the obvious boundary conditions: $f(0) = 0$, since the gambler is bankrupt himself, and $f(a + b) = 1$, since then the opponent is automatically bankrupt.

We discuss the solution of (4) in the following section.

FINITE DIFFERENCE OPERATORS

To see the connection between recurrence relations and difference equations, we must introduce the finite difference operators Δ and E.

The operator Δ : Let $f(n)$ denote a sequence of real numbers defined for n, a positive integer or zero. We define

$$\Delta f(n) = f(n + 1) - f(n). \qquad (5)$$

Clearly,

$$\Delta^2 f(n) = \Delta[\Delta f(n)] = \Delta f(n + 1) - \Delta f(n) \qquad (6)$$

$$= [f(n + 2) - f(n + 1)] - [f(n + 1) - f(n)]$$

$$= f(n + 2) - 2 f(n + 1) + f(n).$$

We call $\Delta f(n)$ the first difference of the function $f(n)$. Similarly $\Delta^2 f(n)$ is the second difference of $f(n)$. It is obvious that we may define differences of any order by using the relation $\Delta^{k+1} f(n) = \Delta[\Delta^k f(n)]$.

A difference equation is any equation involving differences such as

$$\Delta f(n) = n$$

or

$$\Delta^2 f(n) + 2\Delta f(n) = a^n .$$

Using the definitions (5) and (6), we may change these difference equations to the following recurrence relations:

$$f(n + 1) - f(n) = n$$

and

$$f(n + 2) - f(n) = a^n.$$

The converse procedure, that of changing a recurrence relation into a difference equation, is also possible. From (5)

$$f(n + 1) = f(n) + \Delta f(n) , \tag{7}$$

and from (6), by use of (7),

$$f(n + 2) = f(n) + 2\Delta f(n) + \Delta^2 f(n). \tag{8}$$

Using (7) and (8), we may change the recurrence relation (4) to the following difference equation:

$$\Delta^2 f(n) = 0 ,$$

which also follows directly from (6), on substituting (4) in the form obtained by replacing n by $n + 1$.

The solution of $\Delta^2 f(n) = 0$ is easy to find. If $g(n) = \Delta f(n)$, then

$$\Delta[\Delta f(n)] = \Delta g(n) = 0 ,$$

and clearly $g(n)$ must be constant for all values of n; therefore

$$\Delta f(n) = c_1 ,$$

where c_1 is a constant. The solution of this is

$$f(n) = c_1 n + c_2,$$

where c_2 is also a constant. Since $f(0) = 0$ and $f(a + b) = 1$, the constants c_2 and c_1 are determined to be zero and $(a + b)^{-1}$ respectively. Thus

$$f(n) = n(a + b)^{-1}$$

and, finally, $f(a) = a(a + b)^{-1}$ is the solution of the problem of the gambler's ruin.

The operator E: The task of expressing $f(n + k)$ in terms of $f(n)$ and its first k differences, which we have illustrated in (7) and (8), can be greatly simplified by the use of the operator E defined as follows:

$$Ef(n) = f(n + 1) , \tag{9}$$

such that $E^2 f(n) = E[Ef(n)] = Ef(n+1) = f(n+2)$, etc. From definition (5),

$$\Delta f = Ef - f = (E - 1)f.$$

Dropping the function f, we write this as an operational relation,

$$\Delta = (E - 1); \qquad (10)$$

then,

$$\Delta^2 = \Delta(E - 1) = (E - 1)^2 = E^2 - 2E + 1,$$

which gives (6) when applied to $f(n)$. Similarly, we obtain

$$\Delta^k = (E - 1)^k \qquad (11)$$

for all integral values of k.

From (10) we have

$$E = 1 + \Delta,$$

and thus

$$E^k = (1 + \Delta)^k = 1 + \binom{k}{1}\Delta + \binom{k}{2}\Delta^2 + \cdots + \Delta^k, \qquad (12)$$

for all integral values of k. Since

$$E^k f(n) = f(n + k),$$

formula (12) gives

$$f(n + k) = f(n) + \binom{k}{1}\Delta f(n) + \binom{k}{2}\Delta^2 f(n) + \cdots + \Delta^k f(n), \qquad (13)$$

a formula expressing $f(n + k)$ in terms of $f(n)$ and its first k differences. This formula can also be used for interpolation. For example, if we put $k = 1/2$ in (13), we get the infinite series

$$f\left(n + \frac{1}{2}\right) = f(n) + \frac{1}{2}\Delta f(n) - \frac{1}{8}\Delta^2 f(n) + \frac{1}{16}\Delta^3 f(n) \cdots.$$

Of course, the extension of (13) to nonintegral values of k is not valid a priori and must be justified in the particular cases in which it is used.

CONNECTION BETWEEN DIFFERENCES AND DERIVATIVES

Suppose that $f(x)$ is a continuously differentiable function of x. Instead of taking differences with length one, as we did in (5) and (6), let us take differences of length h:

$$\Delta_h f(x) = f(x + h) - f(x).$$

Then,

$$\lim_{h \to 0} \frac{1}{h}\Delta_h f(x) = \lim_{h \to 0} \frac{f(x+h) - f(x)}{h} = \frac{df(x)}{dx} = f'(x). \tag{14}$$

Similarly, if $f(x)$ has k continuous derivatives, we find the following connection between kth differences and kth derivatives:

$$\lim_{h \to 0} \frac{\Delta_h^k f(x)}{h^k} = \frac{d^k f(x)}{dx^k} = f^{(k)}(x). \tag{15}$$

If we put $h = \Delta_h x$, the reason for the standard notation for derivatives becomes clear.

The formulas (14) and (15) express the kth derivative as the limit of kth differences as the value of h goes to zero. It is also possible to express derivatives in terms of differences for a fixed value of h. To do so, we introduce the operator E_h as follows:

$$E_h f(x) = f(x+h).$$

Clearly,

$$E_h = 1 + \Delta_h.$$

For convenience we shall drop the subscript h unless it is needed.

If $f(x)$ is an analytic function of x, then by Taylor's theorem

$$f(x+h) = f(x) + hf'(x) + \frac{h^2}{2!} f''(x) + \frac{h^3}{3!} f'''(x) + \cdots.$$

Let D denote differentiation; thus $Df(x) = f'(x)$. Then Taylor's theorem may be written as

$$Ef(x) = \left[1 + hD + \frac{h^2 D^2}{2!} + \frac{h^3 D^3}{3!} + \cdots\right]f(x) = e^{hD} f(x).$$

Suppressing $f(x)$, we get the operational relation

$$E = e^{hD} = 1 + \Delta. \tag{16}$$

From (16) we find that

$$hD = \log(1 + \Delta) = \Delta - \frac{\Delta^2}{2} + \frac{\Delta^3}{3} - \cdots;$$

applying this to $f(x)$, we obtain the desired formula for the derivative of $f(x)$ in terms of differences, namely,

$$f'(x) = h^{-1}\left[\Delta_h f(x) - \frac{1}{2}\Delta_h^2 f(x) + \frac{1}{3}\Delta_h^3 f(x) - \cdots\right]. \tag{17}$$

The inverse of (17), that is, the expression for $\Delta_h f(x)$ in terms of derivatives of $f(x)$, is given directly by (16):

$$\Delta f = (e^{hD} - 1)f = \left(hD + \frac{h^2}{2}D^2 + \cdots\right)f(x).$$

INVERSE OF A DIFFERENCE OPERATOR

Suppose we are given a function $g(n)$ such that

$$f(n + 1) - f(n) = \Delta f(n) = g(n)$$

for all values of n. (Hereafter, unless otherwise stated, we assume differences of length one.) How can we find $f(n)$? Formally, we may write

$$f(n) = \Delta^{-1} g(n) ,$$

but this is meaningless until Δ^{-1} has been interpreted. To do so, let us return to the definition of difference. The definition gives

$$f(n) - f(n - 1) = g(n - 1)$$

$$f(n - 1) - f(n - 2) = g(n - 2)$$

$$\vdots$$

$$f(1) - f(0) = g(0) .$$

Adding these equations together, we find that

$$f(n) - f(0) = \sum_{k=0}^{n-1} g(k) = \Delta^{-1} g(n) - f(0) . \tag{18}$$

Thus the inverse of differencing is summing. This corresponds to the fact that the inverse of differentiating is integrating. Notice that in summation, as in integration, the result is uniquely defined except for the additive constant $f(0)$.

Just as we could express differences in terms of derivatives, we can also express summation in terms of integration. The easiest way to obtain such an expression is by means of the operational formula (16). We have

$$\Delta = e^D - 1 ;$$

therefore

$$\Delta^{-1} = (e^D - 1)^{-1} \tag{19}$$

Subsequently, we discuss how the right-hand side can be expanded into a power series in D plus a term D^{-1} corresponding to integration, and we use the result to get the Euler-Maclaurin sum formula.

SIMPLE RECURRENCE RELATIONS

The first recurrence relation we discussed, that given by (2), was solved simply by iteration and repeated multiplication. The method used in getting (3) was very similar to that used in getting (18). In fact, if we put

$$f(n) = \log I_n ,$$

relation (2) becomes

$$f(n) - f(n - 2) = \log \frac{n - 1}{n} \, ,$$

which is exactly of the form that led to (18).

The original method used in solving (2) can be extended to a method for solving systems of recurrence relations. Consider the following sequence of fractions:

$$\frac{0}{1}, \frac{1}{1}, \frac{1}{2}, \frac{2}{3}, \frac{3}{5}, \frac{5}{8}, \cdots, \frac{p_n}{q_n}, \cdots$$

where

$$\begin{aligned} p_n &= q_{n-1} \\ q_n &= q_{n-1} + p_{n-1} \, . \end{aligned} \tag{20}$$

We wish to express p_n and q_n as explicit functions of n.

We introduce a vector X_n whose two components are p_n and q_n, and rewrite (20) as

$$X_n = \begin{pmatrix} p_n \\ q_n \end{pmatrix} = \begin{pmatrix} 0 & 1 \\ 1 & 1 \end{pmatrix} \begin{pmatrix} p_{n-1} \\ q_{n-1} \end{pmatrix} = M X_{n-1}, \tag{21}$$

where X_0 is determined from the fact that $p_0 = 0$, $q_0 = 1$. Since the matrix M is independent of n, we use (21) repeatedly for n, $n-1$, $\ldots$, 2, 1, to obtain

$$X_n = M[M X_{n-2}] = \cdots = M^n X_0, \tag{22}$$

or

$$\begin{pmatrix} p_n \\ q_n \end{pmatrix} = M^n \begin{pmatrix} p_0 \\ q_0 \end{pmatrix} = M^n \begin{pmatrix} 0 \\ 1 \end{pmatrix}.$$

Formula (22) would give an explicit formula for p_n and q_n if an explicit formula for M^n were known.

To obtain M^n, explicitly, we could transform it to a diagonal form by a similarity transformation. This means we would determine a diagonal matrix D and a nonsingular matrix P such that

$$M = PDP^{-1} \, .$$

Then

$$M^n = (PDP^{-1})(PDP^{-1}) \cdots (PDP^{-1}) = PD^n P^{-1} \, .$$

But since D is diagonal,

$$D = \begin{pmatrix} \lambda_1 & 0 \\ 0 & \lambda_2 \end{pmatrix}, \qquad D^n = \begin{pmatrix} \lambda_1^n & 0 \\ 0 & \lambda_2^n \end{pmatrix},$$

then we would obtain

$$M^n = P \begin{pmatrix} \lambda_1^n & 0 \\ 0 & \lambda_2^n \end{pmatrix} P^{-1}.$$

Since P and P^{-1} are independent of n, this would give the desired formula for M^n once we had determined D and P.

However, there is a simpler way of evaluating M^n. Consider the characteristic equation of M, namely,

$$c(\lambda) = \begin{vmatrix} 0 - \lambda & 1 \\ 1 & 1 - \lambda \end{vmatrix} = \lambda^2 - \lambda - 1 = 0,$$

whose roots (the eigenvalues) are $\lambda_1 = (1 + \sqrt{5})/2$ and $\lambda_2 = (1 - \sqrt{5})/2$. By a well-known theorem, every matrix satisfies its characteristic equation with unity replaced by the identity matrix $I = \begin{pmatrix} 1 & 0 \\ 0 & 1 \end{pmatrix}$. In the case under consideration we have $M^2 - M - I = 0$, as may be verified from the following computations:

$$M^2 = \begin{pmatrix} 1 & 1 \\ 1 & 2 \end{pmatrix}; \quad M^2 - M - I = \begin{pmatrix} 1 & 1 \\ 1 & 2 \end{pmatrix} - \begin{pmatrix} 0 & 1 \\ 1 & 1 \end{pmatrix} - \begin{pmatrix} 1 & 0 \\ 0 & 1 \end{pmatrix} = \begin{pmatrix} 0 & 0 \\ 0 & 0 \end{pmatrix}.$$

From the characteristic equation, we have

$$M^2 = M + I \, ;$$

multiplying by M and using this equation again gives

$$M^3 = M^2 + M = 2M + I.$$

Similarly, successive multiplication by M would express $M^4, M^5, \ldots, M^n$ as linear functions of M. Let us therefore put

$$M^n = \alpha M + \beta$$

and try to find α and β.

If λ^n is divided by $\lambda^2 - \lambda - 1$, the result would be a quotient $q(\lambda)$ and a remainder $r(\lambda)$ of the first degree in λ. We may write

$$\lambda^n = (\lambda^2 - \lambda - 1) q(\lambda) + r(\lambda), \tag{23}$$

where $r(\lambda) = \gamma \lambda + \delta$, say; we determine γ and δ by using the remainder theorem. Substituting the eigenvalues λ_1 and λ_2 into (23), we obtain

$$\lambda_1^n = \gamma \lambda_1 + \delta, \qquad \lambda_2^n = \gamma \lambda_2 + \delta \, ;$$

therefore

$$\gamma = \frac{\lambda_1^n - \lambda_2^n}{\lambda_1 - \lambda_2}, \tag{24}$$

$$\delta = \frac{\lambda_1 \lambda_2^n - \lambda_2 \lambda_1^n}{\lambda_1 - \lambda_2} = -\lambda_1 \lambda_2 \frac{\lambda_1^{n-1} - \lambda_2^{n-1}}{\lambda_1 - \lambda_2}$$

Since (23) is an algebraic identity, the same result must hold if M is substituted for λ:

$$M^n = (M^2 - M - I)q(M) + \gamma M + \delta I = \gamma M + \delta I$$

where the final form follows because M satisfies its characteristic equation; thus γ and δ correspond to the numbers α and β we sought.

We have

$$M^n = \gamma \begin{pmatrix} 0 & 1 \\ 1 & 1 \end{pmatrix} + \delta \begin{pmatrix} 1 & 0 \\ 0 & 1 \end{pmatrix} = \begin{pmatrix} \delta & \gamma \\ \gamma & \delta + \gamma \end{pmatrix},$$

and from (22)

$$\begin{pmatrix} p_n \\ q_n \end{pmatrix} = \begin{pmatrix} \delta & \gamma \\ \gamma & \delta + \gamma \end{pmatrix} \begin{pmatrix} 0 \\ 1 \end{pmatrix} = \begin{pmatrix} \gamma \\ \delta + \gamma \end{pmatrix}.$$

Using (24), we find that

$$p_n = \frac{\lambda_1^n - \lambda_2^n}{\sqrt{5}}, \qquad q_n = \frac{\lambda_1^{n+1} - \lambda_2^{n+1}}{\sqrt{5}}.$$

It is easy to verify that these will indeed give the numerators and denominators for the fractions we started with.

The above example illustrates a general method for treating recurrence relations between k quantities. If we introduce a k-component vector X_n, the recurrence relation becomes

$$X_n = M_n X_{n-1},$$

and then by use of this relation for $n, n-1, n-2, \ldots, 2, 1$, we find that

$$X_n = M_n M_{n-1} \cdots M_1 X_0.$$

If the matrices $M_n, M_{n-1}, \ldots, M_1$ are such that

$$M_n = M_{n-1} = \cdots = M_1 = M,$$

where M is a constant matrix, then

$$X_n = M^n X_0.$$

The matrix M^n is again evaluated by using the characteristic equation $c(M) = 0$ of the matrix. Dividing λ^n by $c(\lambda)$, we get a quotient $q(\lambda)$ and a remainder

$$r(\lambda) = \alpha_1 \lambda^{k-1} + \alpha_2 \lambda^{k-2} + \cdots + \alpha_k.$$

Since $c(M) = 0$, this implies that

$$M^n = q(M)c(M) + r(M) = r(M).$$

To find $r(\lambda)$, we substitute the eigenvalues $\lambda = \lambda_i$, in the algebraic identity

$$\lambda^n = q(\lambda)c(\lambda) + r(\lambda),\qquad(25)$$

to obtain

$$\lambda_i^n = r(\lambda_i)\qquad(26)$$

for all eigenvalues of M.

If the eigenvalues of M are distinct, (26) gives k linearly independent equations for the k unknown coefficients $\alpha_1, \alpha_2, \ldots, \alpha_k$ of $r(\lambda)$. Clearly, the solution of these equations will determine $r(\lambda)$ and thus M^n. Even if the eigenvalues of M are not all distinct, $r(\lambda)$ can be determined. Suppose, for example, that λ_1 is a multiple eigenvalue of M and therefore a multiple root of $c(\lambda) = 0$. This means that $c'(\lambda)$ and perhaps higher derivatives of $c(\lambda)$ vanish for $\lambda = \lambda_1$. The number of derivatives that vanish depends on the multiplicity of λ_1. Returning to the algebraic identity (25), we find by differentiation

$$n\lambda^{n-1} = q'(\lambda)c(\lambda) + q(\lambda)c'(\lambda) + r'(\lambda).$$

Putting $\lambda = \lambda_1$ gives

$$n\lambda_1^{n-1} = r'(\lambda_1).$$

By taking higher derivatives, we may get other relations involving the coefficients of $r(\lambda)$. It is clear that in this way we shall always get k linearly independent relations for the k coefficients of $r(\lambda)$.

4.2 DIFFERENCE EQUATIONS OF ORDER HIGHER THAN ONE

The method we have discussed can also be used to solve linear difference equations of higher order. Consider the second-order equation (4):

$$f(n+1) = 2f(n) - f(n-1).$$

Put $g(n) = f(n-1)$, and replace the arguments by subscripts; then we may write

$$f_{n+1} = 2f_n - g_n,$$

$$g_{n+1} = f_n.$$

This is a first-order recurrence system which may be solved as before.

We wish to discuss a simpler, more operational, technique of solving linear difference equations with constant coefficients. This technique will be very similar to the Laplace transform method for solving linear differential equations with constant coefficients. The essence of the method consists in considering the numbers $f_0, f_1, \ldots, f_n, \ldots$ as the components of an infinite-dimensional vector φ and using an operator G (a sort of inverse to the operator E) such that

$$G\varphi = G \begin{pmatrix} f_0 \\ f_1 \\ f_2 \\ \vdots \\ f_n \\ \vdots \end{pmatrix} = \begin{pmatrix} 0 \\ f_0 \\ f_1 \\ \vdots \\ f_{n-1} \\ \vdots \end{pmatrix} \qquad (27)$$

We shall illustrate the method by solving the equation

$$f_{n+2} - 4f_{n+1} + 3f_n = 0 ,$$

or

$$f_n - 4f_{n-1} + 3f_{n-2} = 0 . \qquad (28)$$

Notice this latter equation is valid only for $n \geq 2$. Using φ and G as defined before, we write (28) as

$$\varphi - 4G\varphi + 3G^2\varphi = \alpha , \qquad (29)$$

where α is an infinite-dimensional vector all of whose components are zero, except the first and second. The first component of α corresponds to $n = 0$ and is equal to f_0; the second component of α corresponds to $n = 1$ and is equal to $f_1 - 4f_0$.

The formal solution of (29) is obtained by division and separation into partial fractions. We find

$$\varphi = \frac{1}{1 - 4G + 3G^2} \alpha = \frac{1}{2} \left[\frac{3}{1-3G} - \frac{1}{1-G} \right] \alpha$$

$$= \left[\frac{3}{2} \sum 3^n G^n - \frac{1}{2} \sum 1^n G^n \right] \alpha .$$

Comparing the *nth* components of both sides, we see that

$$f_n = a3^n + b1^n, \qquad (30)$$

where a and b are scalars, whose values depend upon f_0 and f_1.

Notice that the solution (30) can be found directly from (28). Assume $f_n = \lambda^n$; then substitution in (28) gives

$$\lambda^2 - 4\lambda + 3 = 0$$

whose roots are $\lambda = 3$ and $\lambda = 1$. Since (28) is a second-order difference equation, its solution must depend upon two arbitrary scalars; thus (30) is obtained.

Let us apply the operational method to equation (4):

$$f(n + 1) = 2f(n) - f(n - 1).$$

We write it as

$$f_n = 2f_{n-1} - f_{n-2},$$

which implies

$$\varphi - 2G\varphi + G^2\varphi = \alpha$$

where the first component of α is f_0, the second is $f_1 - 2f_0$, and all the others are zero. Now

$$\varphi = (1 - G)^{-2}\alpha = \sum nG^{n-1}\alpha.$$

Comparing the nth components, we find

$$f_n = c_1 n + c_2, \tag{31}$$

where c_1 and c_2 are scalars whose values depend on those of f_0 and f_1.

In the problem of the gambler's ruin, we had to solve (4) with the conditions $f(0) = 0$ and $f(a + b) = 1$. Since the general solution of (4) is given by (31), we see that

$$0 = c_2, \qquad 1 = c_1(a + b);$$

therefore

$$f(n) = (a + b)^{-1}n.$$

Note that if we try to obtain (31) by putting $f(n) = \lambda^n$ in (4), we find that $\lambda^2 - 2\lambda + 1 = 0$. This equation has two *equal* roots $\lambda = 1$. One solution of (4) will be $c_2 1^n$, but another solution is needed. To see what it should be, consider a similar differential equation, namely,

$$y'' - 2a y' + a^2 y = 0.$$

The solution of this is found by substituting $y = e^{\lambda x}$ (compare this with substituting $f_n = \lambda^n$). We obtain

$$\lambda^2 - 2a\lambda + a^2 = 0,$$

or $\lambda = a$ is a double root. Since the roots are equal, we know that one solution of the differential equation is e^{ax} and the other, which can be obtained by differentiation with respect to a, is xe^{ax}. In the same way, the characteristic equation for the difference equation

$$f(n + 2) - 2af(n + 1) + a^2 f(n) = 0 \tag{32}$$

is $\lambda^2 - 2a\lambda + a^2 = 0$. One solution of this is a^n and the other, which can be obtained by differentiation, is na^{n-1}; consequently, the complete solution of (32) is

$$f(n) = c_1 na^{n-1} + c_2 a^n.$$

For $a = 1$, this reduces to (31).

NONHOMOGENEOUS DIFFERENCE EQUATIONS

Consider the nonhomogeneous difference equation

$$f(n + 2) - 4f(n + 1) + 3f(n) = a^n. \tag{33}$$

As in the corresponding theory for differential equations, the general solution of (33) can be written as the sum of a particular solution $P(n)$ of (33) and the complementary solution $c(n)$, where $c(n)$ satisfies the homogeneous difference equation

$$c(n + 2) - 4c(n + 1) + 3c(n) = 0.$$

Using the method of the preceding section, we can show that $c(n) = \alpha 3^n + \beta 1^n$, where α and β are arbitrary constants.

To find a solution $P(n)$ which satisfies

$$P(n + 2) - 4P(n + 1) + 3P(n) = a^n,$$

we rewrite the equation as

$$P(n) = \frac{1}{E^2 - 4E + 3} a^n$$

where E is the operator such that $Ef(n) = f(n + 1)$. Since $Ea^n = a^{n+1}$, and

$$(E^2 - 4E + 3)a^n = (a^2 - 4a + 3)a^n,$$

we obtain

$$P(n) = \frac{1}{E^2 - 4E + 3} a^n = \frac{1}{E^2 - 4E + 3} \frac{E^2 - 4E + 3}{a^2 - 4a + 3} a^n = \frac{a^n}{a^2 - 4a + 3}.$$

Similarly, consider

$$(E^2 - 4E + 4)f(n) = n^2.$$

The general solution is

$$f(n) = c(n) + P(n),$$

where $c(n)$ is a solution of

$$(E^2 - 4E + 4)c(n) = 0$$

and $P(n)$ is a particular solution of

$$(E^2 - 4E + 4)P(n) = n^2.$$

By the methods of the preceding section we find that

$$c(n) = \alpha 2^n + \beta n 2^n.$$

To find $P(n)$ we write

$$P(n) = \frac{1}{E^2 - 4E + 4} n^2$$

And use the fact that $E = 1 + \Delta$. Thus

$$P(n) = \frac{1}{1 - 2\Delta + \Delta^2} n^2.$$

Expanding the fraction in a power series in Δ, we get

$$P(n) = [1 + 2\Delta + 3\Delta^2 + 0(\Delta^3)]n^2 = n^2 + 2(2n + 1) + 3 \cdot 2$$

$$= n^2 + 4n + 8,$$

since the third and higher differences of n^2 are all zero.

It is clear that this method for finding the particular solution can be used whenever the right side is a combination of polynomials in n and exponentials in n. We give a final illustration. Consider the equation

$$(E^2 + 1)f(n) = n a^n.$$

The complementary solution $c(n)$ is a solution of

$$(E^2 + 1)c(n) = 0.$$

The method of the preceding section shows that $c(n)$ is a linear combination of i^n and $(-i)^n$, that is, of $e^{\pi i n/2}$ and $e^{-\pi i n/2}$. However, by taking real and imaginary parts of these exponentials, we may write

$$c(n) = \alpha \cos \frac{n\pi}{2} + \beta \sin \frac{n\pi}{2}.$$

The particular solution $P(n)$ will be given by

$$P(n) = \frac{1}{E^2 + 1} \, n\,a^n.$$ (34)

Since

$$E(n\,a^n) = (n + 1)a^{n+1} = a(n\,a^n) + a\cdot a^n,$$

and

$$E^2(n\,a^n) = E[a(n\,a^n) + a\cdot a^n] = a^2(n\,a^n) + 2a^2\cdot a^n,$$

we have

$$(E^2 + 1)n\,a^n = (a^2 + 1)n\,a^n + 2a^2 a^n.$$

Solving for $n\,a^n$, we get

$$n\,a^n = \frac{E^2 + 1}{a^2 + 1}\, n\,a^n - \frac{2a^2}{a^2 + 1}\,a^n.$$

Using this result in (34), we find

$$P(n) = \frac{1}{E^2 + 1}\,\frac{E^2 + 1}{a^2 + 1}\,n\,a^n - \frac{1}{E^2 + 1}\,\frac{2a^2}{a^2 + 1}\,a^n$$

$$= \frac{n\,a^n}{a^2 + 1} - \frac{2a^2}{a^2 + 1}\,\frac{1}{E^2 + 1}\,a^n.$$

Since $(E^2 + 1)a^n = a^n(a^2 + 1)$, we conclude that

$$P(n) = \frac{n\,a^n}{a^2 + 1} - \frac{2a^2}{(a^2 + 1)^2}\,a^n.$$

4.3 THE METHOD OF THE GENERATING FUNCTION

An important method for solving linear differential equations is that of the Laplace transform. A similar method, also due to Laplace, for solving linear difference equations is that of the generating function. We shall illustrate the application of this method in several examples.

Consider the generalized problem of the gambler's ruin. Suppose the gambler starts with a dollars and plays against an opponent who has b dollars. Suppose the probability of the gambler winning one dollar is p, and the probability of losing one dollar is $1 - p = q$. Let $f(n)$ be the probability of the gambler losing all his money when he has n dollars and when, consequently, his opponent has $a + b - n$ dollars. We now have

$$f(n) = p f(n+1) + q f(n-1) ,$$ (35)

where $f(0) = 1$ and $f(a+b) = 0$.

We introduce a generating function $u(t)$ such that

$$\sum_{0}^{\infty} f(n)t^n = u(t).$$

This is very similar to the Laplace transform method in differential equations where, in order to solve the equation

$$y'' + y = 0 ,$$

we introduce the Laplace transform

$$\int_{0}^{\infty} e^{-tx}y(x)\, dx.$$

We multiply (35) by t^n and sum over all values of $n \geq 1$; thus

$$u(t) - f(0) - \sum_{1}^{\infty} f(n)t^n = p\sum_{1}^{\infty} f(n+1)t^n + q\sum_{1}^{\infty} f(n-1)t^n.$$ (36)

Since

$$t\sum_{1}^{\infty} f(n+1)t^n = \sum_{2}^{\infty} f(n)t^n = u(t) - f(0) - tf(1)$$

and

$$\sum_{1}^{\infty} f(n-1)t^n = t\sum_{0}^{\infty} f(n)t^n = tu(t),$$

we may write (36) as

$$u(t) - f(0) = p\frac{u(t) - f(0) - tf(1)}{t} + qtu(t).$$

Solving this for $u(t)$, we find

$$u(t) = \frac{pf(1)t - f(0)(t-p)}{qt^2 - t + p} = \frac{p - t + ptf(1)}{(1-t)(p-qt)}$$ (37)

where we used $f(0) = 1$, and $p + q = 1$.

We now have an expression for the generating function, but what is needed is an expression for $f(n)$. It will be obtained by expanding the right-hand side of (37) in a power series in t. Before expanding, we separate into partial fractions.

We have

$$u(t) = \left[1 + \frac{f(1) - 1}{1 - Q}\right] \frac{1}{1 - t} - \left[\frac{f(1) - 1}{1 - Q}\right] \frac{1}{1 - Qt} , \quad Q = \frac{q}{p} .$$

Since $u(t) = \sum f(n)t^n$, $(1 - t)^{-1} = \sum t^n$, and $(1 - Qt)^{-1} = \sum Q^n t^n$, we equate coefficients of t^n to obtain

$$f(n) = 1 + \left\{\frac{f(1) - 1}{1 - Q}\right\} (1 - Q^n) .$$

Using the fact that $f(a + b) = 0$, we eliminate the factor in braces to obtain

$$f(n) = 1 - \frac{1 - \left(\frac{q}{p}\right)^n}{1 - \left(\frac{q}{p}\right)^{a+b}} .$$

LINEAR DIFFERENCE EQUATIONS WITH NONCONSTANT COEFFICIENTS

The method of the generating function may also be applied to difference equations with nonconstant coefficients. Consider the equation

$$f(n + 2) = nf(n + 1) + nf(n). \tag{38}$$

Again, put

$$\sum_0^\infty f(n)t^n = u(t) . \tag{39}$$

Multiply (38) by t^n and sum for all values of $n \geq 0$. We get

$$\sum_0^\infty t^n f(n + 2) = \sum_0^\infty nt^n f(n + 1) + \sum_0^\infty nf(n)t^n . \tag{40}$$

To take into account the factors of n in the coefficients of the right-hand side, we differentiate (39):

$$u'(t) = \sum_0^\infty nf(n)t^{n-1} = \sum_0^\infty (n + 1)f(n + 1)t^n ,$$

from which we get

$$\sum_0^\infty nt^n f(n + 1) = u'(t) - \frac{u(t) - u(0)}{t} , \quad u(0) = f(0) , \tag{41}$$

as well as

$$\sum_{0}^{\infty} nf(n)t^n = tu'(t) .$$ (42)

Since

$$\sum_{0}^{\infty} f(n+2)t^n = \frac{u(t) - f(0) - tf(1)}{t^2} ,$$ (43)

we can change (40) into a differential equation for $u(t)$ by using (41), (42), and (43). Instead of solving this equation, we shall discuss a different approach which leads to the same result.

METHOD OF INTEGRAL TRANSFORM

To solve (38), we assume that $f(n)$ may be represented by the following integral:

$$f(n) = \int_{\alpha}^{\beta} t^{n-1} v(t) \, dt .$$ (44)

Here the function $v(t)$ and the limits α and β are unknown and must be determined. From (44),

$$nf(n) = \int_{\alpha}^{\beta} n t^{n-1} v(t) \, dt ,$$

and, integrating by parts to get rid of the n's, we find

$$nf(n) = t^n v(t) \Big|_{\alpha}^{\beta} - \int_{\alpha}^{\beta} t^n v'(t) \, dt .$$

Similarly,

$$nf(n+1) = \int_{\alpha}^{\beta} n t^n v(t) \, dt = t^n (tv) \Big|_{\alpha}^{\beta} - \int_{\alpha}^{\beta} t^n (tv)' \, dt .$$

Combining these results and using the fact that

$$f(n+2) = \int_{\alpha}^{\beta} t^{n+1} v(t) \, dt ,$$

we find that equation (38) implies that

$$\int_{\alpha}^{\beta} t^n (tv + tv' + v + v') \, dt - t^n v(1 + t) \Big|_{\alpha}^{\beta} = 0 .$$

To satisfy this equation we assume

$$(t + 1)v + (t + 1)v' = 0 \tag{45}$$

and also

$$t^n v(t)(1 + t)\Big|_\alpha^\beta = 0. \tag{46}$$

The solution of (45) is $v(t) = ce^{-t}$. To satisfy the resulting form of the boundary condition (46),

$$t^n e^{-t}(1 + t)\Big|_\alpha^\beta = 0,$$

we must choose α and β equal to any two of the three values -1, 0, ∞; any choice of two of these will give a solution of (38). For example, if we take $\alpha = 0$, $\beta = \infty$, then

$$f(n) = \int_0^\infty t^{n-1} e^{-t}\, dt = (n - 1)! \,, \tag{47}$$

which satisfies (38) since $(n + 1)! = n(n)! + n(n - 1)! = (n + 1)n!$. Another linearly independent solution is given by putting $\alpha = -1$, $\beta = 0$; for this choice

$$f(n) = \int_{-1}^0 t^{n-1} e^{-t}\, dt. \tag{48}$$

Since (38) is a second-order linear difference equation, the general solution will be a linear combination of (47) and (48).

LINEAR FIRST-ORDER EQUATIONS

The method used to solve first-order linear differential equations with variable coefficients can be generalized to solve first-order linear difference equations. Consider, first, the homogeneous equation

$$f(n + 1) = p_n f(n).$$

Replacing n by $n - 1$, we get

$$f(n) = p_{n-1} f(n - 1)$$
$$\vdots$$
$$f(1) = p_0 f(0) = p_0 C.$$

Then

$$f(n + 1) = p_n p_{n-1} \cdots p_1 p_0 f(0) = C \prod_0^n p_j. \tag{49}$$

Consider now the nonhomogeneous equation

$$f(n + 1) = p_n f(n) + r_n, \tag{50}$$

where p_n and r_n are given functions of n. We assume the solution of (50) is of the same form as (49) but with C a function of n; thus

$$f(n) = C(n) \prod_0^{n-1} p_j .$$

Substituting this in (50), we get

$$C(n+1) \prod_0^n p_j = C(n) p_n \prod^{n-1} p_j + r_n$$

or

$$C(n+1) = C(n) + r_n \left(\prod_0^n p_j \right)^{-1} = C(n) + t_n ,$$

say. Since

$$C(n) = C(n-1) + t_{n-1}$$
$$\vdots$$
$$C(1) = C(0) + t_0 ,$$

we find that

$$C(n+1) = C(0) + \sum_0^n t_k .$$

Finally,

$$C(n+1) = C(0) + \sum_0^n r_k \left(\prod_0^k p_j \right)^{-1} . \tag{51}$$

4.4 SUMMATION BY PARTS

In formula (51) it is required to sum a series of the form

$$\sum_k a_k b_k , \tag{52}$$

where $a_k = r_k$ and $b_k = \left(\prod_0^k p_j \right)^{-1}$. We shall present a method, known as *summation by parts*, which may change the form (52) to a more convenient one.

The method of summation by parts is completely analogous to that of integration by parts. In integration by parts, we have

$$\int u \, dv = uv - \int v \, du .$$

In (52) we have a sum of a product, but nothing that corresponds to dv. However, let us write b_k as a difference:

$$b_k = B_k - B_{k-1}$$

$$b_{k-1} \doteq B_{k-1} - B_{k-2}$$

$$b_1 \doteq B_1 - B_0.$$

Note that

$$B_k - B_0 = b_1 + \cdots + b_k. \tag{53}$$

We define $B_0 = b_0$, and $B_{-1} = 0$, and write

$$\sum_0^n a_k b_k = \sum_0^n a_k (B_k - B_{k-1}) \tag{54}$$

$$= \sum_0^n a_k B_k - \sum_0^n a_k B_{k-1}$$

$$= \sum_0^n a_k B_k - \sum_{-1}^{n-1} a_{k+1} B_k.$$

Combining the terms in (54), we get

$$\sum_0^n a_k b_k = \sum_0^{n-1} (a_k - a_{k+1}) B_k + a_n B_n - a_0 B_{-1}.$$

To emphasize the analogy with integration by parts, we write this result as

$$\sum_0^n a_k b_k = a_n B_n - a_0 B_{-1} - \sum_0^{n-1} B_k (a_{k+1} - a_k). \tag{55}$$

As an illustration of the use of this formula, consider the sum

$$\sum_0^n k r^k.$$

Since we know how to evaluate $\sum r^k$, let

$$B_k = 1 + r + \cdots + r^k = \frac{r^{k+1} - 1}{r - 1}.$$

Then, using (55), we get

$$\sum_0^n k r^k = n \frac{r^{n+1} - 1}{r - 1} - \sum_0^{n-1} \frac{r^{k+1} - 1}{r - 1}$$

$$= n \frac{r^{n+1} - 1}{r - 1} + \frac{n}{r - 1} - \frac{1}{r - 1} \frac{r(r^n - 1)}{r - 1},$$

which may also be obtained from

$$r \frac{d}{dr} \sum_1^n r^k .$$

GENERALIZATION OF SUMMATION BY PARTS

Consider the infinite series

$$S = \sum_0^\infty a_n b_n x^n .$$

Suppose we know the sum

$$\sum_0^\infty b_n x^n = f(x).$$

We shall show how to evaluate S. Put

$$a_n = E^n a_0 = (1 + \Delta)^n a_0 ;$$

then

$$S = \left[\sum b_n x^n (1 + \Delta)^n \right] a_0$$

$$= f(x + x \Delta) a_0 .$$

Let us expand this in powers of Δ by Taylor's theorem. We find

$$S = \left[\sum \frac{x^n \Delta^n}{n!} \frac{d^n}{dx^n} f(x) \right] a_0 \tag{56}$$

$$= \sum_0^\infty \Delta^n a_0 \frac{x^n}{n!} \frac{d^n}{dx^n} f(x).$$

As an illustration of the use of (56) we shall obtain Euler's transformation for alternating series. If we put $b_n = (-1)^n$, then $f(x) = (1 + x)^{-1}$. From the first form of S and from (56) we get

$$S = \sum (-1)^n a_n x^n = \sum \Delta^k a_0 (-1)^k \frac{x^k}{(1 + x)^{k+1}} .$$

If we put $\delta a_0 = a_0 - a_1 = -\Delta a_0$ and similarly use $\delta^k = (-1)^k \Delta^k$, we obtain *Euler's transformation*, namely,

$$\sum (-1)^n a_n x^n = \sum \frac{x^k}{(1 + x)^{k+1}} \delta^k a_0 .$$

For $x = 1$, we get

$$\sum (-1)^n a_n = \sum \frac{\delta^k a_0}{2^{k+1}} .$$

In many cases the right-hand side of this equation is more rapidly convergent than the left-hand side.

EULER-MACLAURIN FORMULA

Another useful summation formula, known as the Euler-Maclaurin formula, replaces sums by integrals. Consider the sum

$$f(1) + \cdots + f(n) = F(n).$$

Given the sequence $f(n)$, how can we find $F(n)$? Using the operational notations, we see that

$$F(n + 1) - F(n) = (E - 1)F(n) = f(n + 1) = Ef(n);$$

therefore

$$F(n) = \frac{E}{E - 1} f(n).$$

However, from Taylor's theorem [see development for (16)],

$$Ef(x) = f(x + 1) = f(x) + f'(x) + \frac{1}{2!} f''(x) + \cdots$$

$$= \left(1 + D + \frac{D^2}{2!} + \frac{D^3}{3!} + \cdots\right)f = e^D f.$$

Symbolically, then, $E = e^D$. Thus

$$F(n) = \frac{e^D}{e^D - 1} f(n) = \left(1 + \frac{1}{e^D - 1}\right) f(n).\qquad(57)$$

We would like to expand $(e^D - 1)^{-1}$ in a power series in D but this is impossible because the function has a pole at $D = 0$. This difficulty may be avoided by working with the function $D(e^D - 1)^{-1}$, which is regular at $D = 0$. In fact,

$$\frac{D}{e^D - 1} = 1 - \frac{D}{2} + O(D^2).$$

Note, however, that

$$\frac{D}{2} + \frac{D}{e^D - 1} = \frac{D}{2} \frac{e^D + 1}{e^D - 1} = \frac{D}{2} \coth \frac{D}{2}.$$

Since this latter function is an even function of D, its power series can contain only even powers of D. We put

$$\frac{D}{2} + \frac{D}{e^D - 1} = 1 + \sum_1^\infty (-1)^{n-1} B_n \frac{D^{2n}}{(2n)!} = 1 + \frac{B_1 D^2}{2!} - \frac{B_2 D^4}{4!} + \cdots\qquad(58)$$

where B_n denote the Bernoulli numbers, $B_1 = \frac{1}{6}$, $B_2 = \frac{1}{30}$, $B_3 = \frac{1}{42}$, etc.

From (58) we get

$$\frac{1}{e^D - 1} = D^{-1} - \frac{1}{2} + \sum_1^\infty (-1)^n B_n \frac{D^{2n-1}}{(2n)!},$$

where D^{-1} means integration. Using this in (57), we obtain the *Euler-Maclaurin formula*:

$$F(n) = \int f(n)dn + \frac{f(n)}{2} + \sum_1^\infty (-1)^k \frac{B_k}{(2k)!} \frac{d^{2k-1}}{dn^{2k-1}} f(n).$$

More generally,

$$f(a) + f(a+1) + \cdots + f(b) = \int_a^b f(\xi)d\xi + \frac{f(a) + f(b)}{2} \qquad (59)$$

$$+ \frac{B_1}{2!}[f'(b) - f'(a)]$$

$$- \frac{B_2}{4!}[f'''(b) - f'''(a)] + \cdots.$$

In most cases the series on the right-hand side does not converge, but even then it provides useful asymptotic expansions.

4.5 SPECIAL METHODS

Just as in the theory of differential equations, there are many difference equations which can be solved only by special tricks. We shall discuss a few examples in order to illustrate some useful special methods.

We begin with the difference equation

$$f(n+2) - 3nf(n+1) + 2n(n-1)f(n) = 0, \qquad (60)$$

whose coefficients vary with n. When we considered first-order equations we showed that a method of variation of constants could solve the nonhomogeneous equation. We shall try a similar device here.

Since the coefficient of $f(n+1)$ has a factor n and the coefficient of $f(n)$ has a factor $n(n-1)$, we consider the equation

$$g(n+2) = ng(n+1).$$

The solution of this is $g(n) = c(n-2)!$ This suggests that we try the substitution $f(n) = c(n)(n-2)!$ in (60). We obtain the equation

$$c(n+2) - 3c(n+1) + 2c(n) = 0, \qquad (61)$$

an equation with constant coefficients which can be solved by the methods discussed previously.

The general solution of (61) is

$$c(n) = \alpha \cdot 2^n + \beta \cdot 1^n;$$

therefore, the general solution of (60) is

$$f(n) = (n-2)!\,(\alpha 2^n + \beta).$$

As another illustration, we consider the problem of the misdirected letters. Suppose that n letters are placed at random in n envelopes. What is the probability that no letter is placed in the correct envelope? This problem is equivalent to the following: suppose the integers 1 to n are written down at random. What is the probability that no integer coincides with its position in the ordered sequences 1 to n?

Instead of working with probabilities, we shall work with the $n!$ permutations of the integers 1 to n. Let $f(n)$ be the number of permutations with no coincidence; then $nf(n-1)$ is the number of permutations with one coincidence, $\binom{n}{2} f(n-2)$ the number with two coincidences, etc. Adding together the number of permutations with every possible number of coincidences, we get

$$n! = f(n) + nf(n-1) + \binom{n}{2} f(n-2) + \cdots + \binom{n}{n-1} f(1) + 1. \qquad (62)$$

The last term on the right-hand side is that of permutations with complete coincidence, namely one.

Equation (62) is a recurrence relation of a completely different type from those we have discussed because the order of the equation is not constant but varies with n. To analyze (62) we introduce an infinite-dimensional vector F whose components are $f(1), f(2), f(3), \ldots$. Then equation (62) implies that

$$
\begin{pmatrix}
1 & 0 & 0 & \cdot & \cdot & \cdot & \cdot & \cdot & \cdot \\
2 & 1 & 0 & \cdot & \cdot & \cdot & \cdot & \cdot & \cdot \\
3 & 3 & 1 & \cdot & \cdot & \cdot & \cdot & \cdot & \cdot \\
\cdot & \cdot & \cdot & \cdot & \cdot & \cdot & \cdot & \cdot & \cdot \\
\binom{n}{n-1} & \binom{n}{n-2} & \cdot & \cdot & \binom{n}{2} & \binom{n}{1} & 1 & 0 & \cdot \\
\cdot & \cdot & \cdot & \cdot & \cdot & \cdot & \cdot & \cdot & \cdot
\end{pmatrix}
F =
\begin{pmatrix}
1! - 1 \\
2! - 1 \\
3! - 1 \\
\cdot \cdot \cdot \\
n! - 1 \\
\cdot \cdot \cdot
\end{pmatrix}
\qquad (63)
$$

Since the matrix on the left-hand side is a triangular one, it can be inverted step by step. However, we shall use a device similar to that used with equation (60) in order to obtain an explicit solution of (62).

Put $f(n) = n! \, q(n)$ in (62). It becomes

$$1 = q(n) + \frac{q(n-1)}{1!} + \frac{q(n-2)}{2!} + \cdots + \frac{q(1)}{(n-1)!} + \frac{q(0)}{n!} \tag{64}$$

where we have put $q(0) = 1$. Let Q be the infinite-dimensional vector with components $q(0), q(1), q(2), \ldots$, and let A be the infinite-dimensional vector all of whose components are one. Then (64) implies that

$$LQ = A, \tag{65}$$

where

$$L = \begin{pmatrix} 1 & & & & \\ 1/1! & 1 & & & \\ 1/2! & 1/1! & 1 & & \\ 1/3! & 1/2! & 1/1! & 1 & \\ \cdot & \cdot & \cdot & \cdot & \cdot \\ \cdot & \cdot & \cdot & \cdot & \cdot & \cdot \end{pmatrix}.$$

If we use the matrix form of the operator G defined in (27),

$$G = \begin{pmatrix} 0 & & & & \\ 1 & 0 & & & \\ 0 & 1 & 0 & & \\ 0 & 0 & 1 & 0 & \\ \cdot & \cdot & \cdot & \cdot \\ \cdot & \cdot & \cdot & \cdot \end{pmatrix},$$

then it is clear that

$$L = I + \frac{1}{1!} G + \frac{1}{2!} G^2 + \frac{1}{3!} G^3 + \cdots = e^G.$$

The solution of (65) is thus

$$Q = e^{-G} A,$$

which gives

$$q(n) = 1 - \frac{1}{1!} + \frac{1}{2!} - \frac{1}{3!} + \cdots + (-1)^n \frac{1}{n!}. \tag{66}$$

Since the probability of no coincidence is $f(n)/n! = q(n)$, formula (66) gives the required answer.

POISSON PROCESS

We now consider a problem which reduces to the solution of a differential-difference equation. Consider a random process such as the emission of electrons from a hot cathode or the number of neutrons striking a Geiger counter in which any number of events (an electron leaving the cathode or a neutron hitting the counter) can occur in a finite time. Let $f_n(t)$ denote the probability of n events occurring between 0 and t. We shall assume that in a small time interval dt the probability of one event occurring is αdt (where α is independent of time) but that the probability of more than one event occurring in that time is $O(dt)$, that is, the probability goes to zero faster than dt.

Consider $f_n(t + dt)$, the probability of n events in time $t + dt$. The n events may occur in any one of the following mutually exclusive ways:

(a) n events in the time $(0,t)$ and none in time t to $t + dt$,

(b) $n - 1$ events in the time $(0,t)$ and one in the time t to $t + dt$,

(c) $n - 2$ events in the time $(0,t)$ and two in the time t to $t + dt$, etc.

Since the probability of no events in the time interval dt is $1 - \alpha dt - O(dt)$, we see that the probability of case (a) is $f_n(t)[1 - \alpha dt - O(dt)]$; the probability of case (b) is $f_{n-1}(t)\alpha dt$; and the probability of case (c) is $O(dt)$. Therefore,

$$f_n(t + dt) = f_n(t)[1 - \alpha dt - O(dt)] + f_{n-1}(t)\alpha dt + O(dt),$$

or

$$\frac{f_n(t + dt) - f_n(t)}{dt} = -\alpha[f_n(t) - f_{n-1}(t)] + O(1).$$

When dt approaches zero, this equation becomes

$$\frac{d}{dt} f_n(t) = -\alpha[f_n(t) - f_{n-1}(t)] \qquad n \geq 1$$

$$\frac{df_0(t)}{dt} = -\alpha f_0(t).$$

(67)

Note that at $t = 0$, $f_n(0) = 0$ for $n > 0$ but $f_0(0) = 1$.

To solve (67) we introduce an infinite-dimensional vector $F(t)$ whose components are the functions $f_0(t), f_1(t), \ldots$. Equation (67) implies

$$\frac{dF}{dt} = -\alpha F + \alpha GF = -\alpha(1 - G)F.$$

Since G is independent of t, we conclude that

$$F = e^{-\alpha t} e^{\alpha G t} F(0) = e^{-\alpha t}\left(1 + \alpha G + \frac{\alpha^2 G^2}{2!} + \ldots\right) F(0).$$

Comparing the nth components of both sides, we get the required result

$$f_n(t) = e^{-\alpha t}\frac{(\alpha t)^n}{n!}.$$

CHAPTER 5

Complex Integration

INTRODUCTION

The method of complex integration is an extremely useful tool in the evaluation of definite integrals and in the study of analytic functions. We shall discuss these applications, with particular attention to the treatment of integrals containing branch points. We begin with some remarks on analytic functions and Cauchy's theorem. Then we consider the calculus of residues and its application to the evaluation of integrals involving functions with pole singularities. We conclude with a discussion of integrals involving branch points, with applications to the gamma function and the hypergeometric function.

5.1 ANALYTIC FUNCTIONS

Let $z = x + iy$ represent a complex number with real part x and imaginary part y. Consider a complex-valued function

$$f(z) = u(x,y) + iv(x,y) ,$$

where $u(x,y)$ and $v(x,y)$ are the real and imaginary parts, respectively, of $f(z)$. The function $f(z)$ is *differentiable* at $z = z_0$ if

$$\lim_{z \to z_0} \frac{f(z) - f(z_0)}{z - z_0}$$

exists, no matter how z approaches z_0. We say $f(z)$ is *analytic* in a domain D if $f(z)$ is differentiable for every z in D. As is well known, differentiability implies the *Cauchy-Riemann conditions*

$$\frac{\partial u}{\partial x} = \frac{\partial v}{\partial y} , \qquad \frac{\partial v}{\partial x} = - \frac{\partial u}{\partial y} \tag{1}$$

The integral of $f(z)$ along a simple curve C is defined as follows:

$$\int_C f(z)\, dz = \int_C (u + iv)(dx + i\, dy) = \int_C (u\, dx - v\, dy) + i \int_C (v\, dx + u\, dy). \tag{2}$$

Let us recall Gauss's theorem, or Green's theorem in the plane. It states that if $P(x,y)$ and $Q(x,y)$ are continuous and continuously differentiable functions in a simply connected region D bounded by a closed simple curve C, then

$$\int_C (P\, dx + Q\, dy) = \iint_D \left(\frac{\partial Q}{\partial x} - \frac{\partial P}{\partial y} \right) dx\, dy .$$

138

Applying this formula to the integrals on the right-hand side of (2), we find that

$$\int_C (u\,dx - v\,dy) = \int\int_D \left(\frac{\partial v}{\partial x} + \frac{\partial u}{\partial y}\right) dx\,dy = 0$$

by (1), and similarly

$$\int_C (v\,dx + u\,dy) = \int\int_D \left(\frac{\partial u}{\partial x} - \frac{\partial v}{\partial y}\right) dx\,dy = 0.$$

This justifies the *Cauchy-Goursat theorem*: If a function $f(z)$ is single-valued and anlytic in a simply connected region D bounded by a closed curve C and continuous on C_1 then

$$\int_C f(z)\,dz = 0.\qquad\qquad (3)$$

The Green's theorem derivation requires that $f'(z)$ be continuous on C, but Goursat showed that (3) holds without this assumption.

SINGULARITIES

The interesting applications of this theorem occur when $f(z)$ is not analytic at all points of D. A point in D at which $f(z)$ is not analytic is called a *singularity* of $f(z)$.

Suppose $f(z)$ is continuous at C and analytic in D except for a singularity at $z = z_0$. Let C_0 be any curve which is contained in D and which encloses the point z_0. We show that

$$\int_C f(z)\,dz = \int_{C_0} f(z)\,dz.$$

To see this, connect the curve C to the curve C_0 by a straight line, as indicated in Figure 5.1.

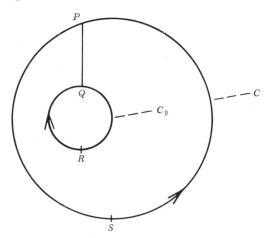

Figure 5.1

Consider the region bounded by the curve $PSPQRQP$, that is, the region between C_0 and C with the line PQ removed. This region is obviously simply connected and contains no singularity of $f(z)$; therefore, by Cauchy's theorem,

$$0 = \int_{PSPQRQP} f(z)\, dz = \left[\int_{PSP} + \int_{PQ} + \int_{QRQ} + \int_{QP} \right] f(z)\, dz.$$

But

$$\left[\int_{PQ} + \int_{QP} \right] f(z)\, dz = 0,$$

since one integral is the negative of the other; consequently,

$$\int_{PSP} = -\int_{QRQ} \; ,$$

which implies that

$$\int_C f(z)\, dz = \int_{C_0} f(z)\, dz \; . \tag{4}$$

In this last result both integrals are to be taken in the positive (counterclockwise) sense.

The result (4) can be easily extended to the case where $f(z)$ has a finite number of singularities in D. Let z_k $(1 \le k \le n)$ be the singular points of $f(z)$ and let C_k be a curve which is contained in D and which encloses z_k but no other singularity of $f(z)$. Then, by a method similar to that used to obtain (4), we find

$$\int_C f(z)\, dz = \sum_{k=1}^{n} \int_{C_k} f(z)\, dz. \tag{5}$$

Suppose, again, that $f(z)$ is analytic in D except at the single point z_0. For simplicity of notation, we may assume $z_0 = 0$. Consider the function $f(z)(z - w)^{-1}$. It is analytic in D except for singularities at $z = w$ and $z = 0$. By (5), we have

$$\int_C \frac{f(z)\, dz}{z - w} = \int_{C_w} \frac{f(z)\, dz}{z - w} + \int_{C_0} \frac{f(z)\, dz}{z - w} \; , \tag{6}$$

where the relationship between C, C_w, and C_0 is illustrated in Figure 5.2. Suppose C_w is a circle of radius ϵ around the point w. Then on C_w we have $z = w + \epsilon\, e^{i\theta}$ and

$$\int_{C_w} \frac{f(z)\, dz}{z - w} = i \int_0^{2\pi} f(w + \epsilon\, e^{i\theta})\, d\theta \; .$$

As ϵ approaches zero, $f(w + \epsilon\, e^{i\theta})$ converges uniformly to $f(w)$; therefore

$$\int_{C_w} \frac{f(z)\, dz}{z - w} = i \lim_{\epsilon \to 0} \int_0^{2\pi} f(w + \epsilon\, e^{i\theta})\, d\theta = if(w) \int_0^{2\pi} d\theta = 2\pi i f(w).$$

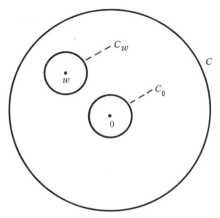

Figure 5.2

Using this in (6), we conclude that

$$f(w) = \frac{1}{2\pi i} \int_C \frac{f(z)}{z-w}\, dz - \frac{1}{2\pi i} \int_{C_0} \frac{f(z)}{z-w}\, dz. \tag{7}$$

On the curve C, $|z| > |w|$ and $|w/z| < 1$; consequently, the series

$$z^{-1} \sum_0^\infty \left(\frac{w}{z}\right)^n$$

converges to $(z-w)^{-1}$. On the curve C_0, however, $|w| > |z|$ and $|z/w| < 1$; consequently,

$$w^{-1} \sum_0^\infty \left(\frac{z}{w}\right)^n$$

converges to $(w-z)^{-1}$. Using these expansions in (7), we find that

$$f(w) = \sum_{-\infty}^\infty a_n w^n, \tag{8}$$

where

$$a_n = \frac{1}{2\pi i} \int_C \frac{f(z)}{z^{n+1}}\, dz, \qquad n = 0, 1, 2, \ldots \tag{9}$$

$$a_{-n} = \frac{1}{2\pi i} \int_{C_0} f(z) z^{n-1}\, dz, \qquad n = 1, 2, 3, \ldots . \tag{10}$$

The series (8) is called the *Laurent* series of the function $f(w)$. We shall use it to investigate the kind of singularity $f(w)$ may have at $w = 0$. Suppose, first, that $f(w)$ is bounded in the neighborhood of $w = 0$. Take C_0 as a circle of radius ϵ around $w = 0$. Then from (10)

$$|a_{-n}| \le \frac{1}{2\pi} \int_0^{2\pi} |f(z)| \epsilon^n\, d\theta .$$

Since the right-hand side of this inequality approaches zero as ϵ approaches zero, we conclude that $a_{-n} = 0$ for $n = 1, 2, 3, \ldots$; therefore, (8) reduces to

$$f(w) = \sum_0^\infty a_n w^n$$

and

$$\lim_{w \to 0} f(w) = a_0 \,.$$

We see that, by putting $f(0) = a_0$, the singularity at $w = 0$ can be "removed." If $f(w)$ is not analytic in the neighborhood of $w = 0$, we make it so by defining $f(0)$ as equal to the limit for $w \to 0$; for example, if $f(w) = (\sin w)/w$, we set $f(0) = 1$.

If $f(w)$ is not bounded in the neighborhood of $w = 0$, an obvious argument shows that not all the negative coefficients a_{-n} can be zero. There are two possibilities: either a finite number of the a_{-n} are not zero, or an infinite number are not zero. If an infinite number are not zero, the point $w = 0$ is called an *essential singularity* of $f(w)$. If a finite number are not zero, let p be the largest value of n for which $a_{-n} \neq 0$. In this case, the point $w = 0$ is called a *pole of order* p for the function $f(w)$ and we have

$$f(w) = \sum_1^p \frac{a_{-n}}{w^n} + \sum_0^\infty a_n w^n .$$

The behavior of an analytic function in the neighborhood of a pole is very simple. The function approaches infinity as z approaches the pole. However, in the neighborhood of an essential singularity the function approaches arbitrarily close to every value and has no limit. This may be illustrated by the behavior of the function $e^{1/z}$ in the neighborhood of $z = 0$, that is, since

$$e^{1/z} = \sum_0^\infty \frac{z^{-n}}{n!} \,,$$

the point $z = 0$ is an essential singularity. As z approaches zero through positive real values, $e^{1/z}$ approaches infinity. As z approaches zero through negative real values, $e^{1/z}$ approaches zero. The function $e^{1/z}$ can be made to approach any desired value as z approaches zero on a suitably chosen sequence of points. For example, if $z = (i2\pi n)^{-1}$, we have

$$e^{1/z} = 1.$$

Suppose $f(z)$ is analytic for every finite value of z and is bounded in the entire z plane. Because $f(z)$ is analytic at $z = 0$, all the negative index coefficients in (8) are zero. Also, from (9),

$$|a_n| \leq \frac{1}{2\pi} \int_C \frac{|f(z)|}{|z|^{n+1}} |dz| .$$

Let C be a circle of radius R and suppose

$$|f(z)| < M$$

for all z. Then

$$|a_n| \le \frac{M}{2\pi} R^{-n} \int_0^{2\pi} d\theta = MR^{-n}. \tag{11}$$

Since R can be made arbitrarily large, this shows that $a_n = 0$ for $n = 1, 2, 3, \ldots$; consequently,

$$f(z) = a_0,$$

a constant. We have thus proved *Liouville's theorem*. A function analytic in the entire complex plane and bounded everywhere must be a constant.

An analytic function with no singularities in the finite part of the complex plane is called an *entire* or *integral* function. Such a function is obviously bounded in any closed region of the complex plane. If it is bounded everywhere, Liouville's theorem shows that it must be a constant; consequently, a nonconstant entire function must be unbounded at infinity and thus have a singularity there. The point at infinity can be mapped into the origin by the transformation $z = 1/t$. Expand the function $f(1/t)$ into a Laurent series around the origin. If the series has only a finite number p of negative powers of t, we say $f(z)$ has a pole of order p at infinity; for this case,

$$f(z) = \sum_1^p a_{-n} z^n + \sum_0^\infty \frac{u_n}{z^n}.$$

If the Laurent series has an infinite number of negative powers of t, we say $f(z)$ has an essential singularity at infinity.

RESIDUES

We have shown in (5) that if the analytic function $f(z)$ has singularities $z_k (1 \le k \le n)$ in D, then

$$\int_C f(z)\, dz = \sum_1^n \int_{C_k} f(z)\, dz.$$

Suppose that the point z_k is a pole of order N of $f(z)$; then the Laurent's series of $f(z)$ in the neighborhood of z_k is

$$f(z) = \sum_1^N a_{-j}(z - z_k)^{-j} + g(z - z_k), \tag{12}$$

where $g(z - z_k)$ is analytic in the neighborhood of z_k. If C_k is completely contained in the region in which the expansion given in (12) is valid, we have

$$\int_{C_k} f(z)\, dz = \sum_1^N a_{-j} \int_{C_k} (z - z_k)^{-j}\, dz + \int_{C_k} g(z - z_k)\, dz. \tag{13}$$

By Cauchy's theorem the last integral is zero. To evaluate the other integrals on the right-hand side we take C_k as a circle of radius ρ around the point z_k; thus, on C_k we have $z = z_k + \rho e^{i\theta}$. We find

$$\int_{C_k} (z - z_k)^{-j} dz = i\rho^{1-j} \int_0^{2\pi} e^{2\pi(1-j)} d\theta, = 0, \qquad j > 1$$

$$= 2\pi i, \quad j = 1 \ .$$

With this result, (13) becomes

$$\int_{C_k} f(z)\, dz = 2\pi i a_{-1} \ . \tag{14}$$

The quantity a_{-1} is called the *residue* of the function $f(z)$ at the pole z_k. It can be obtained either by expanding $f(z)$ in a Laurent series around z_k and finding the coefficient of $(z - z_k)^{-1}$, or by evaluating

$$\lim_{z \to z_k} \frac{1}{(N-1)!} \frac{d^{N-1}}{dz^{N-1}} \left[(z - z_k)^N f(z) \right] \ .$$

In particular, if $f(z)$ has a simple pole at $z = z_k$, that is, if $N = 1$, then the residue of $f(z)$ at z_k is

$$a_{-1} = \lim_{z \to z_k} (z - z_k) f(z) \ . \tag{15}$$

Notice that the proof of (14) depends essentially on N being finite. If N were infinite, which would be the case for z_k an essential singularity, we could not justify the interchange

$$\int_{C_k} \sum_1^\infty a_{-j} (z - z_k)^{-j}\, dz = \sum_1^\infty \int_{C_k} a_{-j} (z - z_k)^{-j}\, dz \ ,$$

and thus we could not get (14).

We now state the residue theorem, which follows from (14) and (15):

Residue Theorem: Let C be a closed curve bounding a simply connected region D. Let $f(z)$ be single valued and analytic in D and in C except for a finite number of poles $z_1, z_2, \ldots, z_n$ in D. If the residues of $f(z)$ at these points are $\mathcal{R}_1, \mathcal{R}_2, \ldots, \mathcal{R}_n$, then

$$\frac{1}{2\pi i} \int_C f(z)\, dz = \sum_{k=1}^n \mathcal{R}_k \ . \tag{16}$$

FULL-PERIOD INTEGRALS OF TRIGONOMETRIC FUNCTIONS

The residue theorem may be used to evaluate integrals of the form

$$I = \int_{-\pi}^{+\pi} g(\cos\theta, \sin\theta)\, d\theta. \tag{17}$$

If we put $z = e^{i\theta}$, then

$$I = -i \oint g\left(\frac{z+z^{-1}}{2}, \frac{z-z^{-1}}{2i}\right) \frac{dz}{z},$$

when the integral is taken around the unit circle. If the integrand is an analytic function of z inside the unit circle, (16) may be used to evaluate it. For example, suppose

$$I = \frac{1}{2\pi} \int_{-\pi}^{+\pi} \frac{d\theta}{a+b\cos\theta}, \tag{18}$$

where a and b are real. By the substitution $z = e^{i\theta}$ we get

$$I = \frac{1}{2\pi i} \oint \frac{2\, dz}{bz^2 + 2az + b}. \tag{19}$$

The singularities of the integrand will occur at the points where

$$bz^2 + 2az + b = b(z - r_1)(z - r_2) = 0. \tag{20}$$

If $|a/b| > 1$ then the roots r_1, r_2 of (10) are real. Since $r_1 r_2 = 1$, we see that one root, say r_1, has magnitude less than unity and the other, r_2, has magnitude greater than unity. Thus the integrand in (19) has a simple pole at $z - r_1$, the root inside the unit circle. By the residue theorem we conclude that

$$I = 2 \lim_{z \to r_1} \left[\frac{z - r_1}{bz^2 + 2az + b} \right] = \frac{2}{b(r_1 - r_2)} = \frac{\operatorname{sgn} a}{\sqrt{a^2 - b^2}}. \tag{21}$$

If $|a/b| < 1$, then, since r_1 and r_2 are complex conjugates and $r_1 r_2 = 1$, we must have r_1 and r_2 on the unit circle. Put

$$r_1 = e^{i\theta_1}, \qquad r_2 = e^{-i\theta_1}.$$

The integral I in (19) has no meaning because the unit circle contour goes through two poles of the integrand. Similarly, the integral in (18) has no meaning because the integrand becomes infinite at $\theta = \pm\theta_1$. We may, however, require the *Cauchy principal value* of I, that is,

$$\lim_{\epsilon \to 0} \left[\int_{-\pi}^{-\theta_1-\epsilon} + \int_{\theta_1+\epsilon}^{\theta_1-\epsilon} + \int_{\theta_1+\epsilon}^{\pi} \right] \frac{d\theta/2\pi}{a+b\cos\theta}.$$

The change of variable $z = e^{i\theta}$ brings this again to the form

$$\frac{1}{2\pi i} \int_{C_0} \frac{dz}{bz^2 + 2az + b},$$

but the contour C_0 is now the unit circle omitting two small arcs around the points $e^{i\theta_1}$ and $e^{-i\theta_1}$, as shown in Figure 5.3.

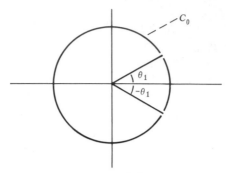

Figure 5.3

Let us close the contour by adding two small semicircular caps of radius ϵ in the neighborhood of the omitted points (see Figure 5.4):

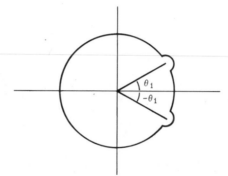

Figure 5.4

We denote these caps by $C_{\pm}$. The value we want is

$$I = \lim_{\epsilon \to 0} \frac{1}{2\pi i} \left[\int_{C_0 + C_+ + C_-} - \int_{C_+} - \int_{C_-} \right] \frac{dz}{bz^2 + 2az + b} . \tag{22}$$

By the residue theorem,

$$\frac{1}{2\pi i} \int_{C_0 + C_+ + C_-} = \mathcal{R}(\theta_1) + \mathcal{R}(-\theta_1) ; \tag{23}$$

that is, the integral over the closed contour equals the residue at θ_1 plus the residue at $-\theta_1$.

Instead of evaluating the integrals over $C_{\pm}$ directly we shall consider a more general integral, namely, the integral over a semicircle surrounding a simple pole. Suppose $f(z)$ has a simple pole at $z = a$ and suppose C_ϵ is a semicircle of radius ϵ with center at $z = a$. Consider

$$I_0 = \lim_{\epsilon \to 0} \frac{1}{2\pi i} \int_{C_\epsilon} f(z)\, dz .$$

Put $z = a + \epsilon e^{i\theta}$ and

$$f(z) = \frac{a_{-1}}{z-a} + g(z),$$

where $g(z)$ is analytic. Then

$$I_0 = \lim_{\epsilon \to 0} \left[\frac{1}{2\pi i} \int_0^\pi a_{-1} i \, d\theta + \frac{\epsilon}{2\pi i} \int_0^\pi g(a + \epsilon e^{i\theta}) e^{i\theta} i \, d\theta \right] = \frac{1}{2} a_{-1}, \tag{24}$$

that is, I_0 equals one-half the residue at the pole.

From (24) we get that

$$\frac{1}{2\pi i} \left[\int_{C_+} + \int_{C_-} \right] = \frac{1}{2} \left[\mathcal{R}(\theta_1) + \mathcal{R}(-\theta_1) \right] ;$$

consequently, by (22) and (23),

$$I = \left(1 - \frac{1}{2} \right) \left[\mathcal{R}(\theta_1) + \mathcal{R}(-\theta_1) \right] = \frac{1}{2b} \left[\frac{1}{r_1 - r_2} + \frac{1}{r_2 - r_1} \right] = 0.$$

One may easily verify that the same result will be obtained for I no matter whether the caps that close C_0 are put both inside, both outside, or one inside and the other outside the circle.

It is clear that the method of this section will succeed if $g(\cos \theta, \sin \theta)$ is a rational function of $\cos \theta$ and $\sin \theta$, because then

$$g\left(\frac{z + z^{-1}}{2}, \frac{z - z^{-1}}{2i} \right)$$

will have only poles as singularities. Consider, however, the following integral representation of the zeroth-order Bessel function:

$$J_0(r) = \frac{1}{2\pi} \int_0^{2\pi} e^{ir\cos\theta} \, d\theta .$$

Again, put $z = e^{i\theta}$. We get

$$J_0(r) = \frac{1}{2\pi i} \oint \left[e^{ir(z + z^{-1})/2} \right] \frac{dz}{z}.$$

This integral, because of the factor $e^{irz^{-1}/2}$, has an essential singularity at the origin; therefore the residue theorem cannot be used. Instead, we expand

$$e^{irz^{-1}/2} = \sum_0^\infty \left(\frac{ir}{2} \right)^n \frac{z^{-n}}{n!} ,$$

and since this series converges uniformly for $|z| > 0$, we may interchange the order of $\int$ and $\sum$:

$$J_0(r) = \frac{1}{2\pi i} \sum_0^\infty \oint e^{irz/2} \frac{(ir/2)^n}{n! z^{n+1}} \, dz .$$

The residue of

$$\frac{e^{irz/2}}{z^{n+1}} = \frac{\sum\limits_{0}^{\infty} \frac{1}{k!} \left(\frac{irz}{2}\right)^{k}}{z^{n+1}}$$

is

$$\left(\frac{ir}{2}\right)^{n} \frac{1}{n!} \, ;$$

therefore by the residue theorem

$$J_0(r) = \sum_{0}^{\infty} \left(\frac{ir}{2}\right)^{n} \frac{1}{n!} \left(\frac{ir}{2}\right)^{n} \frac{1}{n!} = \sum_{0}^{\infty} (-1)^{n} \frac{(r/2)^{2n}}{(n!)^{2}} \, ,$$

the well-known power-series expansion for the Bessel function.

CAUCHY PRINCIPAL VALUE

We digress to indicate an important application of principal value integrals. Suppose

$$w(z) = u(x,y) + iv(x,y)$$

is analytic in the upper half-plane, continuous on the real axis, and is such that the integral

$$\int \left| \frac{w(z)}{z} \right| dz$$

over a large semicircle approaches zero as the radius of the semicircle approaches infinity. Using the residue theorem, we find that

$$w(z) = \frac{1}{2\pi i} \int \frac{w(\zeta) \, d\zeta}{\zeta - z} \tag{25}$$

if the integral is taken over a contour consisting of a large semicircle of radius R and the real axis from $-R$ to R. We now let R approach infinity, so that the integral over the semicircle vanishes, and we get

$$w(z) = \frac{1}{2\pi i} \int_{-\infty}^{\infty} \frac{w(\zeta) d\zeta}{\zeta - z} \, .$$

Put $z = x + iy$ and let y approach zero in this equation. Then

$$w(x) = \lim_{y \to 0} \frac{1}{2\pi i} \int_{-\infty}^{\infty} \frac{w(\zeta) d\zeta}{\zeta - x - iy}$$

$$= \lim_{y \to 0} \frac{1}{2\pi i} \int_{-\infty}^{\infty} \frac{w(\zeta)\{\zeta - x + i y\}}{(\zeta - x)^{2} + y^{2}} \, d\zeta$$

$$= \frac{1}{2\pi i} \, \text{P.V.} \int_{-\infty}^{\infty} \frac{w(\zeta) d\zeta}{\zeta - x} + \frac{i}{2\pi i} \lim_{y \to 0} \int_{-\infty}^{\infty} \frac{yw(\zeta) d\zeta}{(\zeta - x)^{2} + y^{2}} \, ,$$

where P.V. denotes the Cauchy principal value. Using

$$\lim_{y \to 0} \frac{y}{(\zeta-x)^2 + y^2} = \pi \delta(x - \zeta)$$

as in Section 1.3 the last integral gives $\pi w(x)$. Consequently, we obtain

$$w(x) = \frac{1}{\pi i} \, \text{P.V.} \int_{-\infty}^{\infty} \frac{w(\zeta) \, d\zeta}{\zeta - x} \, .$$

Separating into real and imaginary parts, we get finally

$$u(x) = \frac{1}{\pi} \, \text{P.V.} \int_{-\infty}^{\infty} \frac{v(\zeta) \, d\zeta}{\zeta - x} \tag{26}$$

and

$$v(x) = \frac{-1}{\pi} \, \text{P.V.} \int_{-\infty}^{\infty} \frac{u(\zeta) d\zeta}{\zeta - x} \, . \tag{27}$$

Formulas (26) and (27) are examples of Hilbert transforms. Given the real or imaginary part of a function analytic in the upper half-plane, they can be used to determine the imaginary or real part, respectively, and thus the function.

5.2 INTEGRALS OF RATIONAL FUNCTIONS

The methods of contour integration can also be used to evaluate integrals of the form

$$I = \int_{-\infty}^{\infty} g(x) \, dx$$

where $g(x)$ is a rational function of x. Using the symbol $\frown$ to denote integration over a semicircle, the residue theorem states

$$\frac{1}{2\pi i} \left[\int_{-R}^{R} + \int_{\frown} \right] g(z) dz = \sum \mathcal{R}_n$$

is the sum of residues of $g(z)$ in the upper half-plane inside the semicircle of radius R. If as $R \to \infty$, the integral over the semicircle goes to zero, we find

$$I = 2\pi i \sum_{\dagger} \mathcal{R}_n \, ,$$

where the notation is to indicate the sum of residues in the upper half-plane.

As an illustration consider the integral

$$I = \int_{-\infty}^{\infty} \frac{dx}{x^4 + a^4}$$

where $z^4 + a^4 = 0$ has the four roots $z_n = a e^{i\pi(1+2n)/4}$ with $n = 0, 1, 2, 3$; however, only z_0 and z_1 lie in the upper half-plane.

Using L´Hospital's rule, we obtain the residues from

$$\lim_{z \to z_n} \frac{(z - z_n)'}{(z^4 + a^4)'} = \frac{1}{4z_n^3} \ .$$

Thus

$$\frac{1}{2\pi i} \left[\int_{-R}^{R} + \int_{\frown} \right] \frac{dz}{z^4 + a^4} = \left(4a^3 e^{3\pi i/4} \right)^{-1} + \left(4a^3 e^{9\pi i/4} \right)^{-1} = - \frac{i \sin \pi/4}{2a^3} \ .$$

Therefore

$$\int_{-R}^{R} + \int_{\frown} = \frac{\pi \sin(\pi/4)}{a^3} = \frac{\pi}{\sqrt{2} \ a^3} \ .$$

But

$$\left| \int_{\frown} \right| \leq \frac{\pi R}{R^4} \to 0$$

as $R \to \infty$; consequently,

$$I = \frac{\pi}{\sqrt{2} \ a^3} \ .$$

More generally, this method can be used whenever the integrand can be written

$$g(z) = \frac{p(z)}{q(z)} \ ,$$

with $p(z)$ and $q(z)$ as polynomials in z of degrees d_1 and d_2, respectively. In order for the integral

$$\int_{-\infty}^{\infty} \frac{p(z)}{q(z)} \ dz$$

to exist, the difference $d_2 - d_1$ must not be less than two. If that is so, i.e., if $d_2 \geq d_1 + 2$, then the integral over the semicircle

$$\left| \int_{\frown} \frac{p(z)}{q(z)} \ dz \right| \leq \frac{\pi R^{d_1 + 1}}{R^{d_2}} \leq \pi R^{-1} \to 0$$

as $R \to \infty$; consequently

$$\int_{-\infty}^{\infty} \frac{p(z)}{q(z)} \ dz = 2\pi i \sum_+ \mathcal{R}_n \ .$$

Note that we may with equal validity close the contour in the lower half-plane and evaluate the residues there.

INTEGRALS OF EXPONENTIALS MULTIPLIED BY RATIONAL FUNCTIONS

An integral of the form

$$I = \int_{-\infty}^{\infty} e^{ikx}\, \frac{p(x)}{q(x)}\, dx,$$ (28)

where $p(x)$ and $q(x)$ are polynomials of degree d_1 and d_2, respectively, with $d_2 \geq d_1 + 1$, can be evaluated by contour integration. If k is positive (negative), we obtain a complete contour by considering the real axis from $-R$ to R plus a semicircle of radius R in the upper (lower) half-plane. By Cauchy's theorem

$$\frac{1}{2\pi i} \left[\int_{-R}^{R} + \int_{\frown} \right] \frac{p(x)}{q(x)}\, e^{ikx}\, dx = +\sum_{\restriction} \mathcal{R}_n \left(\text{or } -\sum_{\restriction} \mathcal{R}_n \right),$$

where the right-hand side equals plus (or minus) the sum of residues in the upper (lower) half-plane. The negative sign is needed if k is negative because the poles in the lower half-plane are gone around in the clockwise, or negative direction. We shall show that

$$\lim_{R \to \infty} \int_{\frown} \frac{p(x)}{q(x)} e^{ikx}\, dx = 0;$$ (29)

therefore I equals $\pm 2\pi i$ times the sum of the residues of the integrand in the upper or lower half-plane according as k is positive or negative.

We illustrate this method by evaluating

$$I_1 = \int_{-\infty}^{\infty} \frac{e^{ikx}\, dx}{x^2 + a^2}.$$

The integrand has poles at $x = \pm |a| i$. If k is positive, the residue at the pole $|a| i$ in the upper half-plane is

$$\frac{e^{-k|a|}}{2|a| i};$$

therefore

$$I_1 = \frac{\pi e^{-k|a|}}{|a|}.$$

If k is negative, the residue at the pole $-|a| i$ in the lower half-plane is

$$\frac{e^{k|a|}}{-2|a| i}$$

and then

$$I_1 = \frac{\pi e^{k|a|}}{|a|}.$$

Both cases may be written in the form

$$I_1 = \frac{\pi e^{-|ak|}}{|a|} \;.$$

We must still prove that the limit of the integral over the semicircle goes to zero as the radius goes to infinity. This result will follow from *Jordan's lemma*. Let $f(z)$ be analytic in the region of the upper half-plane outside a neighborhood of the origin. Suppose $f(Re^{i\theta})$ approaches zero uniformly for $0 \le \theta \le \pi$ as R approaches infinity. Then the integral

$$J = \int e^{iz} f(z)\, dz \tag{30}$$

over a semicircle of radius R approaches zero as R approaches infinity.

To prove this, put $z = Re^{i\theta}$ in (30). It becomes

$$J = iR \int_0^\pi e^{-R \sin \theta}\; e^{iR \cos \theta}\; f(Re^{i\theta}) e^{i\theta}\, d\theta.$$

Let M_R equal the maximum value of $|f(Re^{i\theta})|$ for $0 \le \theta \le \pi$; then

$$|J| \le M_R R \int_0^\pi e^{-R \sin \theta}\, d\theta = 2 M_R R \int_0^{\pi/2} e^{-R \sin \theta}\, d\theta. \tag{31}$$

Since $\sin \theta \ge 2\theta/\pi$ for $0 \le \theta \le \pi/2$ (see Section 1.3), we obtain

$$|J| \le 2 M_R R \int_0^{\pi/2} e^{-(2/\pi)R\theta}\, d\theta \le \pi M_R \;.$$

By hypothesis, M_R approaches zero as R approaches infinity; therefore, so does J and the theorem is proved.

From our assumptions about (28), the ratio $p(z)/q(z)$ satisfies the inequality

$$\frac{p(z)}{q(z)} \le \frac{\alpha R^{d_1}}{R^{d_2}} = \frac{\alpha}{R^{d_2 - d_1}} \;, d_2 - d_1 \ge 1$$

for sufficiently large values of R, where α is a constant independent of R. This shows that $p(z)/q(z)$ satisfies the assumptions demanded of $f(z)$ in Jordan's lemma; consequently, Jordan's lemma proves (29).

EVALUATION OF INFINITE SERIES

Consider the infinite series

$$\sum_1^\infty \frac{1}{n^2 + a^2} = S \;. \tag{32}$$

We can obtain the closed form S if we can find a function $g(z)$ such that

$$\frac{1}{2\pi i} \oint \frac{g(z)}{z^2 + a^2}\, dz = S \tag{33}$$

where the contour is suitably chosen. To obtain (33), it is clear that $g(z)$ must be a function having poles at the integers $z = n$ and the residue unity at each such pole. A function having the proper pole is $(\sin \pi z)^{-1}$ but it has the residue $\pi^{-1} \cos \pi n$ at the pole $z = n$. This suggests taking

$$g(z) = \pi \cot \pi z ,$$

and indeed $g(z)$ has poles at $z = n$ with residue unity at each pole. Note that for the contour of Figure 5.5

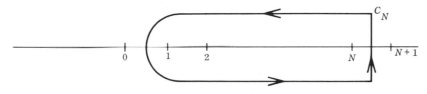

Figure 5.5

we have

$$\frac{1}{2\pi i} \int_{C_N} \frac{\pi \cot \pi z \, dz}{z^2 + a^2} = \sum_1^N \frac{1}{n^2 + a^2} .$$

To evaluate $\mathscr{S}$ consider the square contour S_N formed by the vertical lines $z = \pm(N+1/2) + iy$ and the horizontal lines $z = x \pm (N+1/2)i$. Note that the function

$$\frac{\pi \cot \pi z}{z^2 + a^2}$$

has poles at $z = 0, \pm 1, \ldots, \pm N$ and $\pm ai$ inside S_N; therefore, by Cauchy's theorem

$$\frac{1}{2\pi i} \int_{S_N} \frac{\pi \cot \pi z \, dz}{z^2 + a^2} = \frac{1}{a^2} + 2\sum_1^N \frac{1}{n^2 + a^2} - \frac{\pi \coth \pi a}{a} . \tag{34}$$

We shall show by estimating the integrand on the sides of S_N that the integral in (34) becomes arbitrarily small as N approaches infinity. This will prove that

$$\frac{1}{a^2} + 2\mathscr{S} = \frac{\pi \coth \pi a}{a}$$

or, finally,

$$\mathscr{S} = \frac{\pi \coth \pi a}{2a} - \frac{1}{2a^2} . \tag{35}$$

To estimate (34), note that on S_N

$$\cot \pi\left(N + \frac{1}{2} + iy\right) = i \tanh y$$

and

$$i \cot \pi\left[x + i\left(N + \frac{1}{2}\right)\right] = \frac{e^{-(N+\frac{1}{2})\pi} e^{i\pi x} + e^{(N+\frac{1}{2})\pi} e^{-i\pi x}}{e^{-(N+\frac{1}{2})\pi} e^{i\pi x} - e^{(N+\frac{1}{2})\pi} e^{-i\pi x}}$$

are bounded, and that similar results follow for the other two sides of S_N. These results show that on S_N, for N sufficiently large, the integral in (34) is less than

$$\int_{-\infty}^{\infty} \frac{dy}{\left|\left(N+\frac{1}{2}+iy\right)^2 + a^2\right|} + \int_{-\infty}^{\infty} \frac{dx}{\left|\left[x+i\left(N+\frac{1}{2}\right)\right]^2 + a^2\right|}$$

$$\leq 2\int_{-\infty}^{\infty} \frac{dt}{\left|N+\frac{1}{2}+it\right|^2 - a^2} \leq 2\int \frac{dt}{N^2-a^2+t^2} = \frac{2\pi}{\sqrt{N^2-a^2}}$$

Since this approaches zero as N approaches infinity, (35) is justified.

From this illustration it is clear that series of the form

$$\sum_1^{\infty} f(n)$$

can be evaluated if $f(z)$ is a rational even function of z such that

$$\int_{S_N} |f(z)| \, dz \to 0 \qquad (36)$$

as N approaches infinity.

A similar method may be used to evaluate Fourier series. Consider

$$F = \sum_1^{\infty} \frac{\cos nt}{n^2+a^2}, \quad 0 \leq t \leq \pi.$$

The preceding method cannot be used to evaluate F because (36) is not satisfied. However, we may modify the method by using the integral

$$I_N = \frac{1}{2\pi i} \int_{S_N} \frac{\pi \cos z(\pi-t)}{\left(z^2+a^2\right) \sin \pi z} \, dz,$$

where S_N is the same square contour we have introduced before. Since we assume $0 \leq t \leq \pi$, it follows that the development proceeds essentially as before for $\cot \pi z = \cos \pi z/\sin \pi z$.

The residue of the integrand at $z = \pm n$ is now

$$\frac{1}{n^2+a^2} \frac{\cos n(\pi-t)}{\cos n\pi} = \frac{\cos nt}{n^2+a^2}.$$

By using Cauchy's theorem we see that

$$\lim_{N \to \infty} I_N = 2F + \frac{1}{a^2} - \frac{\pi \cosh a(\pi-t)}{a \sinh \pi a}. \qquad (37)$$

We again estimate I_N on the sides of S_N. On the horizontal sides $z = x \pm i(N+1/2)$,

$$\frac{\cos z\,(\pi - t)}{\sin \pi z} = i\,\frac{e^{iz(\pi-t)} + e^{-iz\,(\pi-t)}}{e^{i\pi z} - e^{-i\pi z}} \sim \mp e^{\pm ixt}\,e^{-t\left(N+\frac{1}{2}\right)}$$

which approaches zero as N approaches infinity. On the vertical sides $z = \pm (N+1/2) + iy$,

$$\left|\frac{\cos z\,(\pi - t)}{\sin \pi z}\right| = \left|\frac{e^{-i\left(N+\frac{1}{2}\right)t}\,e^{-y\,(\pi-t)} - e^{i\left(N+\frac{1}{2}\right)t}\,e^{y\,(\pi-t)}}{e^{-y\pi} + e^{y\pi}}\right| \le 2e^{-t\,|y|}.$$

Since

$$\left|\oint_{S_N} \frac{dz}{z^2 + a^2}\right|$$

approaches zero as N approaches infinity, we conclude that I_N also approaches zero. Therefore (37) implies that

$$F = \frac{\pi \cosh a(\pi - t)}{2a \sinh \pi a} - \frac{1}{2a^2}, \qquad \text{if } 0 \le t \le \pi.$$

5.3 INTEGRALS INVOLVING BRANCH POINTS

Consider the integral

$$I = \int_0^\infty \frac{x^{\alpha-1}\,dx}{1+x}, \qquad 0 < \alpha < 1. \tag{38}$$

If we replace x by $z = x + iy$, we obtain the complex integral

$$I = \int_0^\infty \frac{z^{\alpha-1}\,dz}{1+z}.$$

Because of the factor $z^{\alpha-1}$, the integrand will have a *branch point singularity* at $z = 0$ if $\alpha - 1$ is not a positive integer. To study the behavior of $W = z^{\alpha-1}$ in the neighborhood of $z = 0$, put $z = re^{i\theta}$ and suppose that z goes around a circle C of fixed radius r with θ varying between zero and 2π. We have

$$W = z^{\alpha-1} = r^{\alpha-1}e^{i(\alpha-1)\theta},$$

where $r^{\alpha-1}$ denotes the positive $(\alpha - 1)$th power of r. See Figure 5.6.

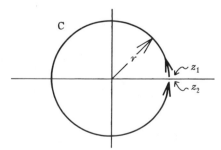

Figure 5.6

For $\theta = 0$, $W(z) = W(z_1)$ has the value $W_0 = r^{\alpha-1}$, but for $\theta = 2\pi$, $W(z) = W(z_2)$ has the value $W_1 = r^{\alpha-1}e^{2\pi i(\alpha-1)}$. Since $\alpha - 1$ is not an integer, W_1 does not equal W_0. As z continues around the circle C, the values of W when z crosses the positive half of the x-axis will be successively

$$W_1 = r^{\alpha-1}e^{2\pi i(\alpha-1)}$$

$$W_2 = r^{\alpha-1}e^{4\pi i(\alpha-1)}$$

$$\vdots$$

and so on. If $\alpha = p/q$, where p and q are integers without common factors, then the values of W will begin to repeat with period q, that is, $W_q = W_0$, $W_{q+1} = W_1$, and so on. However, if α is irrational, then the values of W will form an infinite sequence of distinct values.

It should be remarked that $z = \infty$ is also a branch point for W. To see this, put $z = t^{-1}$ and consider the behavior of W near $t = 0$. We have $W = t^{1-\alpha}$, which obviously has a branch point at $t = 0$. The fact that W has two branch points, namely at $z = 0$ and $z = \infty$, is an illustration of the general theorem that an analytic function of z cannot have only one branch point.

The branch point singularity at $z = 0$ causes difficulty in using Cauchy's theorem to evaluate I because, on any contour containing $z = 0$ in its interior, the integrand will not be single-valued and Cauchy's theorem will not be valid. To avoid this difficulty, we use a contour which does not cross the *branch cut* from 0 to ∞. See Figure 5.7.

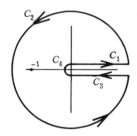

Figure 5.7

Consider the integral

$$I' = \int_{C_1 C_2 C_3 C_4} \frac{z^{\alpha-1}}{1+z}\,dz\ .$$

On and inside the path $C_1 C_2 C_3 C_4$, except for the pole at $z = -1$, the integrand is analytic and single-valued. Thus by Cauchy's theorem I' equals $2\pi i$ times the residue at $z = -1$:

$$I' = 2\pi i\,(e^{\pi i})^{\alpha-1} = -2\pi i e^{\pi i\alpha}\ . \tag{39}$$

On C_1 put $z = x$, and on C_3 put $z = xe^{2\pi i}$. Then

$$I' = \int_0^\infty \frac{x^{\alpha-1}}{1+x}\, dx + \int_{C_2} \frac{z^{\alpha-1}}{1+z}\, dz + e^{2\pi i \alpha} \int_\infty^0 \frac{x^{\alpha-1}}{1+x}\, dx + \int_{C_4} \frac{z^{\alpha-1}\, dz}{1+z} \quad . (40)$$

On C_2 put $z = Re^{i\theta}$, and on C_4 put $z = \epsilon\, e^{i\theta}$. We have

$$\left| \int_{C_2} \right| = R^\alpha \left| \int_0^{2\pi} \frac{ie^{ik\theta}\, d\theta}{1+Re^{i\theta}} \right| \leq 2\pi \frac{R^\alpha}{R-1}$$

which approaches zero as R approaches infinity. For the integral over C_4, we find

$$\left| \int_{C_4} \right| = \epsilon^\alpha \left| \int_0^{2\pi} \frac{ie^{i\alpha\theta}\, d\theta}{1+\epsilon e^{i\theta}} \right| \leq \frac{2\pi\epsilon^\alpha}{1-\epsilon}$$

which also approaches zero as ϵ approaches zero. Consequently, (40) reduces to

$$I' = (1 - e^{2\pi i \alpha})I$$

with I as in (38); then using (39) to eliminate I', we obtain

$$I = \frac{-2\pi i e^{\pi i \alpha}}{1 - e^{2\pi i \alpha}} = \frac{\pi}{\sin \alpha \pi} \quad . \tag{41}$$

For another illustration of how to handle integrals involving branch points, consider the following integral representation of the zeroth-order Bessel function:

$$J_0(r) = \frac{1}{\pi} \int_0^\pi e^{ir \cos\varphi}\, d\varphi \ .$$

Put $\cos\varphi = x$; then

$$J_0(r) = \frac{1}{\pi} \int_{-1}^1 \frac{e^{ixr}\, dx}{\sqrt{1-x^2}} \quad . \tag{42}$$

This formula suggests considering the complex integral whose integrand is

$$\frac{e^{irz}}{(1-z^2)^{1/2}} \ .$$

Since this integrand has branch points at $z = \pm 1$, it is reasonable to connect the branch points by a branch cut along the real axis and to consider the integral

$$J = \frac{1}{2\pi} \int_C \frac{e^{irz}\, dz}{(1-z^2)^{1/2}}$$

where C is any contour enclosing the branch cut. [See Figure 5.8.]

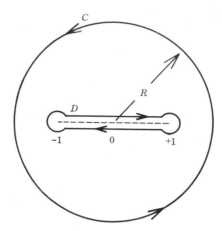

Figure 5.8

The integrand of J is single-valued and analytic in the region bounded by C and the dumbbell-shaped contour D (which we eventually shrink) surrounding the branch cut. Thus we may use Cauchy's theorem to conclude that

$$\left[\int_C + \int_D\right] \frac{e^{irz}\,dz}{(1-z^2)^{1/2}} = 0,$$

and therefore represent J by

$$J = -\frac{1}{2\pi} \int_D \frac{e^{irz}\,dz}{(1-z^2)^{1/2}} . \tag{43}$$

The expression for J given in (43) is not determinate until we specify the meaning of $(1-z^2)^{1/2}$. The square root of a complex quantity can have two values differing by a minus sign, and we must always specify which value we mean. It is not necessary to specify the branch of the square root for all values of z in C. It is sufficient to do it at one point because then by continuity its value at all other points will be determined. Let us suppose C is a circle of radius R and take the point $z = iR$ on C. We have $(1-z^2)^{1/2} = (1+R^2)^{1/2} = \pm\, |\,1+R^2\,|^{1/2}$. We *specify* that for $z = iR$, $(1-z^2)^{1/2} = +\,|\,1+R^2\,|^{1/2}$. Because of this specification, the value of $(1-z^2)^{1/2}$ can now be uniquely determined for all points on C and all points on D by continuity.

Before doing this, we consider the simpler function

$$W = (1+z)^{1/2},$$

Suppose that we specify $W = +1$ for $z = 0$. How is W determined at other values of z? Put

$$1 + z = re^{i\theta}$$

where $z = 0$ corresponds to $r = 1$, $\theta = 0$. We have

$$W = (re^{i\theta})^{1/2} = \pm r^{1/2} e^{i\theta/2} .$$

But since $W = +1$ for $z = 0$, we must have

$$W = r^{1/2} e^{i\theta/2} . \tag{44}$$

This clearly gives a unique value of W as long as θ is restricted to lie between zero and 2π. Of course, if θ increases by 2π, thus indicating that z has made a complete circuit around the branch point -1, the value of W will be multiplied by -1. This fact is typical for the behavior of a function near a square-root branch point.

Formula (44) also has the following interpretation: When the phase θ of z relative to the branch point increases by an amount α, then the phase of W increases by only $\alpha/2$. This simple fact will be very helpful in determining the proper values of $(1 - z^2)^{1/2}$. Notice that

$$t = (1 - z^2)^{1/2} - [(1 + z)(1 - z)]^{1/2} ;$$

consequently, we must take into account the phase of z relative to both branch points at $+1$ and -1. Thus, to find the value of t at $z = 0$ on the upper edge of D, we reason as follows: As z goes from iR (for which we specified $t = +|1+R^2|^{1/2}$) to $z = 0$, the phase of z relative to -1 decreases by $\arctan R$, whereas the phase of z relative to $+1$ increases by the same amount; consequently, the phase of t does not change, and we have $t = +1$ for $z = 0$. However, consider the value of t for $z = 0$ on the lower edge of D. To get from $z = iR$ to this point, we may assume z goes around C in the positive direction to $z = -iR$ and then up the imaginary axis to $z = 0$. In so doing, the phase of z relative to -1 increases by $2\pi - 2\arctan R$ for the path along C and then increases by $\arctan R$ for the path along the imaginary axis, giving a total increase relative to -1 of $2\pi - \arctan R$. The phase of z relative to $+1$ increases by $2 \arctan R$ for the path along C but then decreases by $\arctan R$ for the path along the imaginary axis, thus giving a total increase in phase of $\arctan R$. Since the phase of a product is the sum of the phases, we see that the phase of $(1 - z^2)$ increases by $(2\pi - \arctan R) + \arctan R = 2\pi$; consequently, the phase of t increases by $(1/2)2\pi = \pi$; therefore, for $z = 0$ on the lower edge of D, we have $t = -1$.

A similar argument can be used to determine t at every point in the plane. Put $z = x$ on the upper and lower edges of D. We have $(1 - z^2)^{1/2} = +(1 - x^2)^{1/2}$ on the upper edge and $(1 - z^2)^{1/2} = -(1 - x^2)^{1/2}$ on the lower edge, where $(1 - x^2)^{1/2}$ means the positive real square root. Since the integrals about the little circles going around ± 1 go to zero as $\sqrt{\epsilon}$ when their radius ϵ goes to zero, we find that

$$\int_D \frac{e^{irz} \, dz}{\sqrt{1 - z^2}} = \int_{-1}^{1} \frac{e^{irx} \, dx}{\sqrt{1 - x^2}} - \int_{1}^{-1} \frac{e^{irx} \, dx}{\sqrt{1 - x^2}} = 2\int_{-1}^{1} \frac{e^{irx} \, dx}{\sqrt{1 - x^2}} .$$

From (43) and (42) we conclude that

$$J_0(r) = -J .$$

Suppose we wish to obtain a power series for $J_0(r)$. We may expand e^{irz} in a uniformly convergent power series to obtain

$$J_0(r) = -\frac{1}{2\pi} \sum \frac{(ir)^n}{n!} \int_C \frac{z^n \, dz}{(1 - z^2)^{1/2}} . \tag{45}$$

To evaluate these integrals, put $z = Re^{i\theta}$ and let the radius R of C go to infinity. We have

$$-\frac{1}{2\pi} \int_C \frac{z^n \, dz}{(1 - z^2)^{1/2}} = \lim_{R \to \infty} \frac{1}{2\pi i} \int_0^{2\pi} \frac{R^{n+1} e^{i(n+1)\theta} \, d\theta}{(1 - R^2 e^{2i\theta})^{1/2}} .$$

But

$$(1 - R^2 e^{2i\theta})^{1/2} = Re^{i(\theta - \pi/2)}(1 - R^{-2} e^{-2i\theta})^{1/2} ,$$

where the factor $e^{i(\theta - \pi/2)}$ accounts that for $\theta = \pi/2$, the value of $(1 - R^2 e^{2i\theta})^{1/2}$ is real and positive. Expanding $(1 - R^{-2} e^{-2i\theta})^{1/2}$ in a uniformly convergent binomial series in inverse powers of R, we find that

$$\frac{R^n}{2\pi} \int \frac{e^{ni\theta} \, d\theta}{(1 - R^{-2} e^{-2i\theta})^{1/2}} = \frac{1}{2\pi} \sum (-1)^k R^{n-2k} \binom{-1/2}{k} \int_0^{2\pi} e^{i(n-2k)\theta} \, d\theta$$

$$= \left[(-1)^k \binom{-1/2}{k} \right]_{k=n/2} = \binom{1/2}{n/2}$$

$$= \frac{n!}{\left[\left(\frac{n}{2} \right)! \right]^2 2^{2n}}$$

if n is even, and equals zero if n is odd; here $\binom{\alpha}{m}$ is the binomal coefficient:

$$\binom{\alpha}{m} = \begin{cases} \alpha(\alpha - 1)(\alpha - 2) \ldots (\alpha - m + 1)/m! \text{ for } m = 1, 2, 3, \ldots ; \\ 1 \text{ for } m = 0 \end{cases}$$

From (45), putting $n = 2m$, we conclude that

$$J_0(r) = \sum_0^\infty \frac{(-1)^m}{(m!)^2} \left(\frac{r}{2} \right)^{2m} ,$$

the well-known result obtained earlier by an alternative development.

Integrals involving branch points are closely related to Euler's integrals of both the first and second kind. We consider the Euler integral for the beta function and its representation in terms of the gamma function:

$$B(\alpha,\beta) = \int_0^1 x^{\alpha-1}(1-x)^{\beta-1}\,dx = \frac{\Gamma(\alpha)\Gamma(\beta)}{\Gamma(\alpha+\beta)}; \quad \alpha > 0, \quad \beta > 0, \quad (46)$$

where $\Gamma(z+1) = z\Gamma(z)$, and $\Gamma(n+1) = n!$ for $n = 0, 1, 2, \ldots$. The integrand considered as a function of z, namely $z^{\alpha-1}(1-z)^{\beta-1}$, has obvious branch points at $z = 0$ and $z = 1$ if α and β are not integers. Put $z = t^{-1}$ and the integrand becomes

$$(t-1)^{\beta-1} t^{2-\alpha-\beta}$$

which has a branch point at $t = 0$ if $\alpha+\beta$ is not an integer. Thus, unless $\alpha+\beta$ is an integer, the function $z^{\alpha-1}(1-z)^{\beta-1}$ has three branch points, namely, at $z = 0$, $z = 1$, and $z = \infty$.

When $\alpha+\beta$ is an integer, the integrand has two branch points, and the integral in (46) may be evaluated by a method similar to that used to evaluate the integral in (45). We consider

$$\int_D z^{\alpha-1}(1-z)^{\beta-1}\,dz \qquad (47)$$

where D is the contour shown in Figure 5.9.

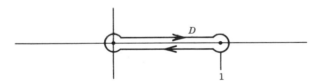

Figure 5.9

Because of the branch points at $z = 0$ and $z = 1$, formula (47) has no meaning until we define which branch of the integrand we shall consider. We define the branch by stating that the function $W = z^{\alpha-1}(1-z)^{\beta-1}$ should be positive real for z on the upper edge of D. As z goes in the direction of increasing x along the upper edge of D and then in a clockwise direction around the circle about $z = 1$ to the lower edge of D, the phase of z relative to $z = 0$ remains constant, but the phase of z relative to $z = 1$ decreases by 2π; consequently, the phase of W decreases by $2\pi(\beta - 1)$. As z goes in the direction of decreasing x along the lower edge of D and then in a clockwise direction around the circle at $z = 0$ to the upper edge of D, the phase of z relative to $z = 0$ decreases by 2π, but the phase of z relative to $z = 1$ remains constant; consequently, the phase of z decreases by $2\pi(\alpha - 1)$. The total change in phase of W as z describes the contour D in the direction indicated by the arrows is therefore a decrease of $2\pi(\alpha+\beta-2)$. If $\alpha+\beta$ is an integer, this change of phase in W is an integral multiple of 2π which leaves the value of W unchanged; consequently, a complete circuit of D does not change the value of W. If $\alpha + \beta$ is not an integer, the change of phase in W will not be an integral multiple of 2π; thus, a complete circuit of D will change the value of W by the factor $\exp[-2\pi i(\alpha+\beta-2)]$. Because of this difference in the values of W we must consider the contour D as not being closed if $\alpha + \beta$ is not an integer.

Suppose that $\alpha + \beta$ is an integer. The contour D may be considered as closed, and the integral around D as the sum of the integrals along the upper and lower edges plus the integrals around the small circles. The integrals around the circle will go to zero as the radii of the circles go to zero. On the horizontal edges of D put $z = x$. Note that we have shown before that the value of W on the lower edge is the value of W on the upper edge multiplied by $\exp[-2\pi i(\beta-1)] = \exp[-2\pi i(\alpha-1)]$. Therefore,

$$\int_D z^{\alpha-1}(1-z)^{\beta-1}\,dz = \left[1 - e^{-2\pi i(\beta-1)}\right] \int_0^1 x^{\alpha-1}(1-x)^{\beta-1}\,dx \qquad (48)$$

$$= \left[1 - e^{-2\pi i(\beta-1)}\right] B(\alpha,\beta).$$

The integral around D can now be evaluated by the method used in evaluating the integrals in (45), that is, by expanding the contour D to a circle of radius R and letting R approach infinity. Put $z = Re^{i\theta}$. For $\theta = 0$, the phase of W will have decreased from its value on the upper edge of D by the amount $\pi(\beta-1)$; therefore

$$W = R^{\alpha-1}e^{i\theta(\alpha-1)}\left[1 - Re^{i\theta}\right]^{\beta-1}$$

$$= R^{\alpha+\beta-2}e^{i\theta(\alpha+\beta-2)}e^{-i\pi(\beta-1)}(1 - R^{-1}e^{-i\theta})^{\beta-1}$$

on the circle of radius R; consequently,

$$\int_D W\,dz = -R^{\alpha+\beta-1}i\int_0^{2\pi} e^{-\pi i(\beta-1)}e^{i\theta(\alpha+\beta-1)}(1 - R^{-1}e^{-i\theta})^{\beta-1}\,d\theta$$

$$= 2\pi i e^{-\pi i(\beta-1)}\binom{\beta-1}{\alpha+\beta-1}(-1)^{\alpha+\beta}\quad.$$

Equating this to (48), and using (46), we get

$$B(\alpha,\beta) = \frac{\pi}{\sin\pi(\beta-1)}\binom{\beta-1}{\alpha+\beta-1}(-1)^{\alpha+\beta} = \frac{\Gamma(\alpha)\Gamma(\beta)}{\Gamma(\alpha+\beta)}\quad.$$

Since $\Gamma(\alpha+\beta) = (\alpha+\beta-1)!$ for $\alpha+\beta = n$, we have

$$\Gamma(\alpha)\Gamma(\beta) = \frac{(-1)^{\alpha+\beta}\pi}{\sin\pi(\beta-1)}(\beta-1)(\beta-2)\cdots(1-\alpha),$$

or, equivalently,

$$\Gamma(\alpha)\Gamma(n-\alpha) = \frac{\pi}{\sin\alpha\pi}(1-\alpha)(2-\alpha)\cdots(n-1-\alpha), \qquad (49)$$

where for $n = 1$ the product on the right-hand side should be replaced by 1.

The result for $n = \alpha + \beta = 1$, i.e.,

$$\Gamma(\alpha)\Gamma(1 - \alpha) = \frac{\pi}{\sin \pi\alpha} \quad , \tag{50}$$

can be obtained from (41) by a simple change of variable. We have from (46) with $\Gamma(\alpha + \beta) = \Gamma(1) = 1$,

$$\Gamma(\alpha)\Gamma(1 - \alpha) = \int_0^1 x^{\alpha - 1}(1 - x)^{-\alpha} \, dx \; .$$

This integrand has branch points at $x = 0$ and $x = 1$. We use a linear transformation which maps 0 into 0 and 1 into ∞, namely,

$$x = \frac{t}{1+t} \; , \qquad dx = \frac{dt}{(1+t)^2} \quad .$$

Then

$$\Gamma(\alpha)\Gamma(1 - \alpha) = \int_0^\infty \left(\frac{t}{1+t}\right)^{\alpha - 1} (1+t)^\alpha \frac{dt}{1+t^2} = \int_0^\infty \frac{t^{\alpha - 1}}{1+t} \, dt = \frac{\pi}{\sin \alpha\pi}$$

by (41).

The counting of branch points and the use of linear transformations enables us to codify a number of important definite integrals. Start with (46):

$$\frac{\Gamma(\alpha)\Gamma(\beta)}{\Gamma(\alpha + \beta)} = \int_0^1 x^{\alpha - 1}(1 - x)^{\beta - 1} \, dx \; . \tag{51}$$

The integrand has three branch points at 0, 1, ∞ and the integral is over a path joining two of them. By the transformation

$$x = \frac{c - b}{b - a} \frac{y - a}{c - y} \; ,$$

we obtain the correspondence

x	0	1	∞
y	a	b	c

Note that c is not between a and b. Making this transformation in (40), we find that

$$\frac{\Gamma(\alpha)\Gamma(\beta)}{\Gamma(\alpha + \beta)} = \frac{(c-b)^\alpha (c-a)^\beta}{(b-a)^{\alpha + \beta - 1}} \int_a^b \frac{(y-a)^{\alpha - 1}(b-y)^{\beta - 1}}{(c-y)^{\alpha + \beta}} \, dy \; . \tag{52}$$

Notice that this integrand has in general only three branch points at a, b, c; the point at infinity is not a branch point. Of course, if any of α, β, or $\alpha + \beta$ is an integer, the integrand will have only two branch points. In that case just as in (49), the evaluation of the integrand will not require gamma functions but only trigonometric functions. A particularly interesting case of (52) is obtained by taking $b = \infty$. In that case the transformation is

$$x = \frac{y-a}{y-c}, \qquad dx = +\frac{a-c}{(y-c)^2}\, dy$$

and we find

$$\frac{\Gamma(\alpha)\Gamma(\beta)}{\Gamma(\alpha+\beta)} = (a-c)^\beta \int_a^\infty \frac{(y-a)^{\alpha-1}}{(y-c)^{\alpha+\beta}}\, dy \; .$$

Notice again that this integrand has three branch points, i.e., a, c, and infinity.

A typical integral with four branch points is

$$G = \int_0^1 x^{\beta-1}(1-x)^{\gamma-\beta-1}(1-ux)^{-\alpha}\, dx, \qquad |u| < 1. \tag{53}$$

The integrand has branch points at $x = 0, 1, u^{-1}$, and ∞. The fact that infinity is a branch point can be seen by putting $x = t^{-1}$ in the integrand. It becomes

$$t^{-(\beta-1)}\frac{(t-1)^{\gamma-\beta-1}}{t^{\gamma-\beta-1}}\frac{(t-u)^{-\alpha}}{t^{-\alpha}} = \frac{(t-1)^{\gamma-\beta-1}(t-u)^{-\alpha}}{t^{\gamma-\alpha-2}}$$

which has a branch point at $t = 0$ unless $\alpha - \gamma$ is an integer. Since $|u| < 1$, we may expand $(1-ux)^{-\alpha}$ by the binomial theorem and integrate term by term. We get

$$G = \int_0^1 x^{\beta-1}(1-x)^{\gamma-\beta-1}\left[1 + \alpha ux + \frac{\alpha(\alpha+1)}{1\cdot 2}u^2 x^2 + \cdots\right] dx \tag{54}$$

$$= \frac{\Gamma(\beta)\Gamma(\gamma-\beta)}{\Gamma(\gamma)}\left[1 + \frac{\alpha\beta}{\gamma}u + \frac{\alpha(\alpha+1)\beta(\beta+1)}{1\cdot 2\cdot\gamma\cdot(\gamma+1)}u^2 + \cdots\right]$$

$$= \frac{\Gamma(\beta)\Gamma(\gamma-\beta)}{\Gamma(\gamma)}F(\alpha,\beta;\gamma;u)$$

where $F(\alpha,\beta;\gamma;u)$ is the hypergeometric function. Note that if $u = 1$, or if either $\gamma-\beta$ or $\gamma-\alpha$ is an integer, the integrand will have only three branch points; consequently, F will be expressible in terms of gamma functions. For example,

$$F(\alpha,\beta;\gamma;1) = \frac{\Gamma(\gamma)\Gamma(\gamma-\alpha-\beta)}{\Gamma(\gamma-\alpha)\Gamma(\gamma-\beta)}\; . \tag{55}$$

5.4 CONTOUR INTEGRAL REPRESENTATIONS OF SOLUTIONS TO THE WAVE EQUATION

The two-dimensional wave equation

$$\left(\frac{\partial^2}{\partial r^2} + \frac{1}{r}\frac{\partial}{\partial r} + \frac{1}{r^2}\frac{\partial^2}{\partial\varphi^2} + k^2\right)u = 0 \tag{56}$$

has as solutions plane waves traveling in an arbitrary direction α. In polar coordinates these plane waves are

$$u = e^{ikr\cos(\varphi-\alpha)}. \tag{57}$$

Since u is a solution of (56) for any value of α, then by superposition the integral

$$\int A(\alpha)e^{ikr\cos(\varphi-\alpha)}\,d\alpha$$

will also be a solution of (56). We first determine the infinite contours for which the integral exists. Put

$$\varphi - \alpha = \beta = \beta_1 + i\beta_2$$

where β_1 and β_2 are real. We have

$$\cos\beta = \cos\beta_1\cosh\beta_2 - i\sin\beta_1\sinh\beta_2;$$

therefore

$$\left| e^{ikr\cos\beta} \right| = e^{kr\sin\beta_1\sinh\beta_2}.$$

For the exponential to go to zero as α goes to ∞, we must require

$$\sin\beta_1\sinh\beta_2 < 0. \tag{58}$$

In order to satisfy this inequality, the path of integration must go to infinity in one of the shaded regions indicated in Figure 5.10.

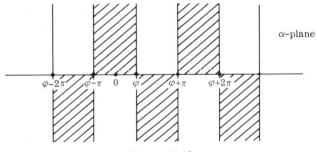

Figure 5.10

Let us try to find solutions of (56) which are of the form

$$u = e^{i\nu\varphi}V(r).$$

By substituting in (56), we find that $V(r)$ must satisfy the equation

$$V'' + \frac{1}{r}V' + \left(k^2 - \frac{\nu^2}{r^2}\right)V = 0, \tag{59}$$

which is Bessel's equation. Let us try to write $V(r)$ as a superposition of plane waves in the form

$$V(r) = \int e^{ikr \cos \alpha} A(\alpha) \, d\alpha \qquad (60)$$

where the path of integration will be specified later. Applying the differential operator in (59) to the integral in (60), we see that we must have

$$\int A e^{ikr \cos \alpha} \left[-k^2 \cos^2 \alpha + \frac{ik \cos \alpha}{r} + k^2 - \frac{\nu^2}{r^2} \right] d\alpha = 0$$

for all r. Since α is independent of r, the bracket is not obviously zero. We try to remove the r terms from the bracket by using integration by parts. First, however, we multiply by r^2 and use $1 - \cos^2 \alpha = \sin^2 \alpha$ to get

$$E = \int A(\alpha) e^{ikr \cos \alpha} \left[-k^2 r^2 \cos^2 \alpha + ikr \cos \alpha + k^2 r^2 - \nu^2 \right] d\alpha$$

$$= -\int A(\alpha) \left[\frac{\partial^2}{\partial \alpha^2} - \nu^2 \right] e^{ikr \cos \alpha} \, d\alpha = 0 \, .$$

Then we use $QP'' = PQ'' + (QP' - Q'P)'$ and integrate by parts to obtain

$$E = (Aikr \sin \alpha + A') e^{ikr \cos \alpha} \; \Big| \; -\int e^{ikr \cos \alpha} (A'' + \nu^2 A) d\alpha \, ,$$

where the first term on the right-hand side must be evaluated at the ends of the path of integration. If

$$A'' + \nu^2 A = 0 \, ,$$

that is, if $A = e^{\pm \nu \alpha i}$, the integral term will disappear. If the path of integration ends in one of the shaded regions of Figure 5.11, the factor $e^{ikr \cos \alpha}$ will go to zero and consequently E will be zero.

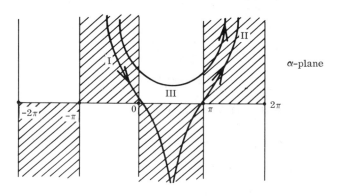

Figure 5.11

We conclude that

$$\int e^{ikr\cos\alpha}\, e^{i\nu\alpha}\, d\alpha$$

over any one of the paths marked I, II, and III above will be a solution of (59). However, (59) is a differential equation of the second order and can have only two independent solutions. Since by Cauchy's theorem

$$\int_I + \int_{II} = \int_{III},$$

because the integrand has no singularities in the finite part of the plane, we see that the linearly independent solutions of (56) can be obtained by taking any two of the three paths I, II, or III of the figure.

Suppose $\nu = n$ is a nonnegative integer, and consider the integral over path III. By Cauchy's theorem, this integral is equal to

$$\left[\int_{-\pi+i\infty}^{-\pi} + \int_{-\pi}^{\pi} + \int_{\pi}^{\pi+i\infty} \right] e^{ikr\cos\alpha}\, e^{in\alpha}\, d\alpha \ .$$

Since the integrand is periodic with period 2π, the integral over the vertical line $-\pi + i\infty$ to $-\pi$ cancels the integral over the vertical line π to $\pi + i\infty$; consequently

$$\int_{-\pi}^{\pi} e^{ikr\cos\alpha}\, e^{in\alpha}\, d\alpha$$

is a solution of (59). We would like a modification that equals a real function of kr. This integral is not a real function of kr because if we put $\beta = \alpha - (\pi/2)$ and use the periodicity to keep the limits the same, we obtain

$$e^{in(\pi/2)} \int_{-\pi}^{\pi} e^{-ikr\sin\beta}\, e^{in\beta}\, d\beta = e^{in(\pi/2)} \int_{-\pi}^{\pi} \cos\,(kr\sin\,\beta - n\beta)\, d\beta,$$

since

$$\int_{-\pi}^{\pi} \sin\,(kr\sin\,\beta - n\beta)\, d\beta = 0$$

because the argument of the sine is an odd function of β. This discussion shows that the integral

$$J_n(kr) = \frac{1}{2\pi} \int_{-\pi}^{\pi} e^{ikr\cos\alpha}\, e^{in\,[\alpha-(\pi/2)]}\, d\alpha \tag{61}$$

will be a real function of kr. The integral represents the Bessel function of order n. The normalization factor $1/2\pi$ was introduced to secure the standard version with $J_0(0) = 1$.

If ν is not an integer, we cannot simplify the integral over path III as we have done above. In this case it is more convenient to use the paths I and II. We define the Hankel functions of order ν as

$$H_\nu^{(1)}(kr) = \frac{1}{\pi} \int_I e^{ikr\cos\alpha} e^{i\nu[\alpha-(\pi/2)]} \, d\alpha, \tag{62}$$

$$H_\nu^{(2)}(kr) = \frac{1}{\pi} \int_{II} e^{ikr\cos\alpha} e^{i\nu[\alpha-(\pi/2)]} \, d\alpha, \tag{63}$$

and the corresponding Bessel function of order ν by the relation

$$H_\nu^{(1)} + H_\nu^{(2)} = 2J_\nu. \tag{64}$$

Consequently,

$$J_\nu(kr) = \frac{1}{2\pi} \int_{III} e^{ikr\cos\alpha} e^{i\nu[\alpha-(\pi/2)]} \, d\alpha \tag{65}$$

reduces to (61) if $\nu = n$, an integer.

The power series expansion for $J_n(kr)$ can be obtained from (61) by expanding $e^{ikr\cos\alpha}$ in a power series in kr and integrating term by term. This method does not work for $J_\nu(kr)$ in (65) because the integrals

$$\int_{III} (\cos\alpha)^m e^{i\nu[\alpha-(\pi/2)]} \, d\alpha$$

do not converge. To get the power series expansion, we must be more subtle. Let us take

$$\int_{III} = \int_{-\pi+\infty}^{-\pi} + \int_{-\pi}^{\pi} + \int_{\pi}^{\pi+\infty} = \int_{C_1} + \int_{C_2} + \int_{C_3}$$

Put $t = e^{-i\alpha}$ in these integrals. We get

$$J_\nu(kr) = \frac{1}{2\pi} \left[\int_{C_1'} + \int_{C_2'} + \int_{C_3'} \right] e^{ikr(t+t^{-1})/2} \, t^{-\nu} \, e^{-i\nu\pi/2} \frac{i \, dt}{t} \tag{66}$$

where C_1', C_2', C_3' are the images of C_1, C_2, C_3, respectively. The path C_1 is defined by $\alpha = -\pi + i\alpha'$ where α' is real; therefore C_1' is defined by $t = e^{+i\pi} e^{\alpha'}$. Similarly, C_3 is defined by $\alpha = \pi + i\alpha'$ and t by $e^{-i\pi} e^{\alpha'}$. Finally, C_2 is defined by $\alpha = \alpha'$, where α' is real, and between $-\pi$ and π; consequently, C_2' is defined by $|t| = 1$. Thus the image of III in the t-plane is as shown in Figure 5.12:

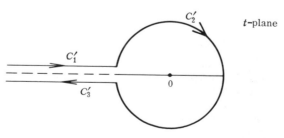

Figure 5.12

At $t = 0$, the integrand in (66) has an essential singularity (because of the term t^{-1} in the exponent) and also a branch point (because of $t^{-\nu}$). Expanding $e^{ikrt^{-1}/2}$ in a power series in kr, we find that

$$J_\nu(kr) = \frac{i}{2\pi} \sum \left(\frac{ikr}{2}\right)^n \frac{1}{n!} \int e^{ikrt/2} e^{-i\nu\pi/2} \frac{dt}{t^{\nu+1+n}} \ . \tag{67}$$

The integral in (67) may be recast as

$$\int \frac{e^{ikrt/2}}{t^{\nu+n+1}} dt = \left(\frac{ikr}{2}\right)^{\nu+n} \int \frac{e^t \, dt}{t^{\nu+n+1}} \ .$$

To evaluate the new version

$$I_{n+\nu} = \int \frac{e^t \, dt}{t^{\nu+n+1}}$$

we notice that integration by parts gives

$$I_{n+\nu} = \frac{1}{\nu + n} I_{n+\nu-1} \tag{68}$$

which is the behavior of $1/\Gamma(z)$. Consider

$$I_{(-\gamma)} = \int \frac{e^t \, dt}{t^{1-\gamma}} \ ,$$

where $0 < \gamma < 1$. First, the integral over C'_2 can be made to go to zero by letting the radius of C'_2 go to zero. Second, because of the branch point at $t = 0$, the value of the integrand along C'_3 is $e^{i\pi(1-\gamma)2}$ times the value of the integrand along C'_1; therefore

$$I_{-\gamma} = \left[e^{i\pi(1-\gamma)2} - 1 \right] \int_0^{-\infty} \frac{e^t \, dt}{t^{1-\gamma}} = -e^{-i\pi(1-\gamma)} \left[e^{i2\pi(1-\gamma)} - 1 \right] \int_0^\infty e^t t^{\gamma-1} dt$$

$$= -i2 (\sin\pi\gamma)\Gamma(\gamma) = -i\frac{2\pi}{\Gamma(1-\gamma)} \ .$$

From (68) and this result we conclude that

$$I_{n+\nu} = \frac{-i2\pi}{\Gamma(n+\nu+1)} \ .$$

Finally, using this in (67) we get

$$J_\nu(kr) = \sum_0^\infty (-1)^n \left(\frac{kr}{2}\right)^{2n+\nu} \frac{1}{n!\,\Gamma(n+\nu+1)} \ .$$

CHAPTER 6

Symbolic Methods

INTRODUCTION

The use of symbolic and operational methods is as old as the invention of calculus. An enormous amount of work and ingenuity was expended in deriving operational techniques for most of the problems in differential and difference calculus. The heyday for this "operationalism" was England in the nineteenth century, and various topics we shall consider can be found in either the 1865 edition of Boole's *Differential Equations* or the 1860 edition of Boole's *Calculus of Finite Differences*.

Even though symbolic and operational methods are very powerful in deriving formulas and extremely useful in systematizing and understanding them, these methods have been neglected by modern mathematicians. The difficulty with the methods is that it is very hard to justify the results obtained. In many cases it is easier to start with the results and justify them directly than to try to justify the operational manipulations involved in obtaining them. However, the study of functional analysis has enabled mathematicians to carry out the justification of operational methods in many important cases. Since functional analysis requires more mathematical background than we wish to assume, we shall of necessity restrict ourselves to mostly formal derivations of important results with very little discussion of their justification.

6.1 DIFFERENTIAL OPERATORS

An operator is defined by giving its domain, that is, the set of functions on which it acts, and its action on functions in its domain. Consider the differential operator D. The domain of D is the set of all differentiable functions $u(x)$. Given $u(x)$, D acts on it as follows: $Du(x) = du(x)/dx$.

If we are given operators D_1, D_2, etc., we may define new operators that are functions of these operators. For example, the operator αD_1 is defined by

$$(\alpha D_1)u(x) = \alpha[Du(x)] ; \tag{a}$$

the operator $D_1 + D_2$ is defined by

$$(D_1 + D_2)u(x) = D_1 u(x) + D_2 u(x); \tag{b}$$

the operator $D_1 D_2$ is defined by

$$(D_1 D_2)u(x) = D_1[D_2 u(x)] . \tag{c}$$

In (c) a problem may arise if the function $D_2 u(x)$ is not in the domain of D_1. We shall ignore such problems. In general D_1 and D_2 are not commutative, i.e., $D_1 D_2 \neq D_2 D_1$, but for the present we will assume commutativity.

From these operators, an algebra may be formed containing such operations as αD, $D_1 + D_2$, $D_1 D_2$, and polynomials such as $P(D) = \alpha_0 + \alpha_1 D + \alpha_2 D^2 + \cdots + \alpha_n D^n$. For convenience we introduce an identity operator I such that

$$I u(x) = u(x).$$

A polynomial in D then becomes $\alpha_0 I + \alpha_1 D + \cdots + \alpha_n D^n$.

Using what we have so far discussed, we can derive Leibnitz's formula for the nth derivative of a product. If D is the differentiation operator, we may write

$$D(uv) = uv' + vu' = (D_1 + D_2)uv \tag{1}$$

where D_1 is an operator that differentiates only u and derivatives of u, and D_2 is an operator that acts similarly on v. Note that D_1 and D_2 commute, that is, $D_1 D_2 (uv) = D_2 D_1 (uv)$. Now

$$D - D_1 + D_2, \qquad D^2 = (D_1 + D_2)(D_1 + D_2) - D_1^2 + 2D_1 D_2 + D_2^2,$$

or in general

$$D^n - D_1^n + \binom{n}{1} D_1^{n-1} D_2 + \cdots + D_2^n. \tag{2}$$

Applying (2) to a product uv, we get

$$D^n(uv) = vu^{(n)} + \binom{n}{1} v' u^{(n-1)} + \cdots + uv^{(n)}, \qquad v^{(n)} = D^n v,$$

which gives us one method of finding the nth derivative of a function (uv).

As an illustration, let $u = e^{ax}$; then since $D^k(e^{ax}) - a^k e^{ax}$, we find that

$$D^n(e^{ax} v) = va^n e^{ax} + \binom{n}{1} v' a^{n-1} e^{ax} + \cdots + e^{ax} v^{(n)}$$

$$= e^{ax} \left[a^n I + \binom{n}{1} a^{n-1} D + \cdots + D^n \right] v$$

$$= e^{ax} [aI + D]^n v$$

$$= e^{ax} (D + a)^n v.$$

Dropping v, we get the *shift rule*:

$$D^n e^{ax} \doteq e^{ax} (D + a)^n. \tag{3}$$

This is an equation between operators. The qualified equality $\doteq$ means here that the same result will be obtained by having each side act on a function in the domain of the operators, that is,

$$(D^n e^{ax}) v = [e^{ax}(D + a)^n] v .$$

For example,

$$D(e^{ax} v) = e^{ax} v' + a e^{ax} v = e^{ax}(D + a) v .$$

Note that D and e^{ax} do not commute.

The shift rule can be derived by repeated application of first-order operations:

$$D^3 e^{ax} = D^2 D e^{ax}$$

$$= D^2 e^{ax}(D + a)$$

$$= D D e^{ax}(D + a)$$

$$= D e^{ax}(D + a)^2$$

$$= e^{ax}(D + a)^3 .$$

Since the shift rule $D^n e^{ax} \doteq e^{ax}(D + a)^n$ is good for all values of n, then by addition the shift rule is valid for polynomials. If $P(D) = \alpha I + \alpha_1 D + \cdots + \alpha_n D^n$, then

$$P(D) e^{ax} \doteq e^{ax} P(D + a) . \tag{4}$$

We wish to use operators to represent integrals as well as derivatives, but there is a slight difficulty. Let us use $\int$ to represent the operation of integration. Then, if we put $D^0 = I$ we have

$$D \int u = u = D^0 u .$$

This suggests that we put $\int = D^{-1}$; but notice that

$$\int Du = \int u' = u + c ,$$

where c is a constant. Therefore

$$D^{-1} D \neq D D^{-1}$$

because

$$u + c \neq D D^{-1} u = u$$

In the next section we will explain a method for remedying this non-commutativity.

What shall be meant by the operator $(D + a)^{-1}$? If $(D + a)^{-1} u = v$, then $u = (D + a) v$; that is, v is a solution of the differential equation $v' + av = u$. Since no initial conditions are given, the solution is not unique. The formulas we shall obtain later will be valid for any solution of the differential equation.

It is easy to see that, as long as we restrict ourselves to polynomials of operators, everything is rigorous and all our results can be proved without much difficulty. The interesting results, however, arise from the use of *infinite* expressions involving operators, but then we must discuss convergence questions. What does an infinite series of operators mean? To what kind of functions does it apply, etc.?

For example, we have shown that if D operates on a product uv, it may be written as $D_1 + D_2$, where D_1 operates on u alone and D_2 operates on v alone. It seems reasonable then to write

$$\frac{1}{D} = \frac{1}{D_1 + D_2} = \frac{1}{D_1\left(1 + \dfrac{D_2}{D_1}\right)} \; ,$$

and to expand the right-hand side in an infinite series:

$$\frac{1}{D} = \frac{1}{D_1} - \frac{D_2}{D_1^2} + \frac{D_2^2}{D_1^3} - \cdots$$

Now, the question is what does this mean? Apply both sides to the product uv and we get

$$\int uv = v \int u - v' \left(\int\right)^2 u + v'' \left(\int\right)^3 u \cdots \tag{5}$$

where $\left(\int\right)^2 u$ means the second integral of u, etc. The questions of what are the constants of integration and what functions can be used in order to get a convergent series on the right-hand side, are questions in functional analysis. We shall not discuss these questions but we remark that if $v(x)$ is a polynomial, the right-hand side of (5) reduces to a finite number of terms, and it then represents an integral of uv.

As an illustration, take $u = e^{ax}$ and $v = x^2$. We get

$$\int e^{ax} x^2 \, dx = \frac{x^2}{a} e^{ax} - \frac{2x}{a^2} e^{ax} + \frac{2}{a^3} e^{ax} \, . \tag{6}$$

This can also be obtained by direct application of the shift rule:

$$\frac{1}{D} e^{ax} \doteq e^{ax} \frac{1}{(D + a)} \, . \tag{7}$$

To verify (7), consider

$$\frac{1}{D} (e^{ax} v) = e^{ax} \frac{1}{(D + a)} v \, ,$$

$$\int (e^{ax} x^2) \, dx = e^{ax} \frac{1}{(D + a)} x^2 \, ;$$

but

$$\frac{1}{a + D} = \frac{1}{a} - \frac{D}{a^2} + \frac{D^2}{a^3} - \frac{D^3}{a^3(D + a)} \, ,$$

and

$$\frac{1}{a + D} x^2 = \frac{x^2}{a} - \frac{2x}{a^2} + \frac{2}{a^3} + 0 .$$

Note that the equation

$$\frac{1}{D + a} x^2 = w$$

can be rewritten as the differential equation $(D + a)w = x^2 = w' + aw$.

Analytic functions of operators may be defined by their power series expansions. The definition of e^{aD} is

$$e^{aD} = I + aD + \frac{a^2 D^2}{2!} + \frac{a^3 D^3}{3!} + \cdots . \qquad (8)$$

To see its use, apply (8) to the function $u(x)$ and get

$$e^{aD} u(x) = u(x) + au'(x) + \frac{a^2}{2!} u''(x) + \cdots .$$

This is the Taylor series for $u(x + a)$ when $u(x)$ is analytic; when $x = 0$, we have the MacLaurin series

$$e^{aD} u(0) = u(0) + au'(0) + \cdots .$$

We now have a definition of the *shift operator*:

$$e^{aD} u(x) \equiv u(x + a) . \qquad (9)$$

These methods can be used to obtain some formulas for numerical evaluation of integrals. We have

$$\int_x^{x+a} u(\xi) d\xi = U(x + a) - U(x)$$

$$= (e^{aD} - 1) U(x)$$

$$= (e^{aD} - 1) \frac{1}{D} u(x) .$$

But

$$\frac{e^{aD} - 1}{D} = a + \frac{a^2 D}{2!} + \frac{a^3 D^2}{3!} + \cdots ;$$

therefore

$$\int_x^{x+a} u \, d\xi = au(x) + \frac{a^2 u'(x)}{2!} + \frac{a^3 u''(x)}{3!} + \cdots . \qquad (10)$$

Equation (10) gives the value of the integral from A to B, as in Figure 6.1,

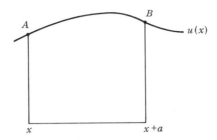

Figure 6.1

in terms of the value of the function and its derivatives at A. We obtain a better approximation if we also use values of $u(x)$ at B. To approximate the integral in terms of values at both A and B, we take the following steps:

$$\frac{e^{aD} - 1}{D} = \frac{e^{aD} - 1}{D} \frac{e^{aD} + 1}{e^{aD} + 1}$$

$$= \frac{1}{D} \frac{e^{aD} - 1}{e^{aD} + 1} \left(e^{aD} + 1\right)$$

$$= \frac{1}{D} \frac{e^{aD/2} - e^{-aD/2}}{e^{aD/2} + e^{-aD/2}} \left(e^{aD} + 1\right)$$

$$= \frac{\tanh aD/2}{D} \left(e^{aD} + 1\right)$$

$$= \left(e^{aD} + 1\right)\left[\frac{aD}{2} - \frac{a^3 D^3}{24} + O(D^5)\right]/D.$$

Applying this to our integral formula, we get

$$\int_x^{x+a} u(\xi) d\xi = (e^{aD} + 1)\left[\frac{au(x)}{2} - \frac{a^3 u''(x)}{24} + O(u^{(4)})\right]$$

(11)

$$= \frac{a}{2}\left[u(x) + u(x + a)\right] - \frac{a^3}{24}\left[u''(x) + u''(x + a)\right] + O(u^{(4)}).$$

This would be exact for a cubic equation since the correction term, which depends on fourth derivatives, would then be zero. Formula (11) is more useful than (10) because the error in (11) depends upon the fourth derivatives, whereas in (10) the error depends upon the magnitude of the third derivatives.

We can obtain a still better formula than (11) by using the expansion for sinh instead of tanh. We proceed as follows:

$$\frac{e^{aD} - 1}{D} = \frac{e^{aD/2} - 1}{D} (e^{aD/2} + 1) = \frac{e^{aD/4} - e^{-aD/4}}{D} \left[e^{3aD/4} + e^{aD/4} \right]$$

$$= \frac{2 \sinh \dfrac{aD}{4}}{D} \left[e^{3aD/4} + e^{aD/4} \right]$$

$$= \frac{2}{D} \left[\frac{aD}{4} + \left(\frac{aD}{4} \right)^3 \frac{1}{3!} + \left(\frac{aD}{4} \right)^5 \frac{1}{5!} + \cdots \right] \left[e^{3aD/4} + e^{aD/4} \right].$$

Therefore,

$$\int_x^{x+a} u(\xi)\,d\xi = \frac{a}{2} \left[u\left(x + \frac{3a}{4} \right) + u\left(x + \frac{a}{4} \right) \right]$$

$$+ \frac{a^3}{192} \left[u''\left(x + \frac{3a}{4} \right) + u''\left(x + \frac{a}{4} \right) \right] + O(u^{(4)}). \tag{12}$$

Note that now the coefficient of the second-derivative term is much smaller than in (11).

6.2 DIFFERENCE OPERATORS

We define

$$\Delta u(x) = u(x + 1) - u(x) \tag{13}$$

and

$$Eu(x) = u(x + 1). \tag{14}$$

Therefore,

$$\Delta u(x) = Eu(x) - u(x)$$

and

$$\Delta = E - 1, \qquad E = 1 + \Delta. \tag{15}$$

We now form an algebra with these operators. We have

$$E^2 u(x) = u(x + 2), \qquad E^n u(x) = u(x + n) \tag{16}$$

and

$$\Delta^2 u(x) = \Delta[u(x + 1) - u(x)]$$

$$= [u(x + 2) - u(x + 1)] - [u(x + 1) - u(x)]$$

$$= u(x + 2) - 2u(x + 1) + u(x).$$

Using the binomial theorem, we get

$$\Delta^n = (E - 1)^n = E^n - \binom{n}{1} E^{n-1} + \cdots + (-1)^n \, 1 \, . \tag{17}$$

Applying this to $u(x)$, we find

$$\Delta^n u(x) = u(x + 1) - \binom{n}{1} u(x + n - 1) + \cdots \text{ (binomial expansion)} \, .$$

When n is an integer we get a finite series and an exact answer. Applying the binomial theorem to $E^n = (1 + \Delta)^n$, we get

$$u(x + n) = u(x) + \binom{n}{1} \Delta u + \binom{n}{2} \Delta^2 u + \cdots + \Delta^n u \, . \tag{18}$$

Again, when n is an integer this formula is exact. When n is not an integer, we obtain an infinite series and questions of convergence arise. A finite number of terms of this series, however, are customarily used to give interpolation formulas. For example, put $n = 1/2$; then

$$u\left(x + \frac{1}{2}\right) = u(x) + \frac{1}{2} \Delta u(x) - \frac{1}{8} \Delta^2 u(x) + \cdots \, .$$

We now investigate the connection between derivative operators and difference operators. Since

$$e^D u(x) = u(x + 1) \, ,$$

we see that

$$e^D = E = 1 + \Delta \, . \tag{19}$$

Taking the logarithm of both sides we get

$$D = \ln (1 + \Delta) = \Delta - \frac{\Delta^2}{2} + \frac{\Delta^3}{3} - \cdots \tag{20}$$

or

$$u'(x) = \Delta u - \frac{\Delta^2 u}{2} + \frac{\Delta^3 u}{3} \cdots \, .$$

This is an expression for the derivative of a function in terms of its differences.

Consider the operator Δ^{-1}. To find its meaning put $u = \Delta^{-1} v$; then

$$v(x) = u(x + 1) - u(x) \, ,$$

$$v(x - 1) = u(x) - u(x - 1) \, ,$$

$$v(x - 2) = u(x - 1) - u(x - 2), \text{ etc.}$$

Adding these, we get

$$\sum v(x) = u(x + 1) \, ;$$

therefore the inverse of the difference operation is summation:

$$\Delta^{-1} = \sum .$$

We can now obtain a formula for a sum in terms of derivatives. We have

$$\sum = \Delta^{-1} = \frac{1}{e^D - 1} = \left[\frac{1}{D + \dfrac{D}{2} + \cdots} - \frac{1}{D} \right] + \frac{1}{D} = \left[\right] + \int ,$$

so that $\sum$ equals $\int$ plus a power series in D. This series gives the well-known Euler-MacLaurin summation formula (see Chapter 4 for further details).

6.3 LAPLACE TRANSFORMS

In this section we shall discuss the theory of the Laplace transform. We shall use operational methods following the ideas (but not the details) of Heaviside and of others. Parts of the present development will be similar to that of Chapter 1, but we shall not use complex integration. We shall treat the topic in a self-contained way instead of as an illustration of a broader subject.

Following Heaviside, we use p (instead of D) to designate the differentiation operator such that $pf(t) = f'(t)$. As we remarked earlier, there is a difficulty in defining the inverse of the operator p. Suppose the inverse operator p^{-1} is defined as the integral operator, that is, $p^{-1}f(t) = \int^t f(\tau)\,d\tau$, then

$$p\,\frac{1}{p}\,f(t) = p \int^t f(\tau)\,d\tau = f(t) ,$$

but

$$\frac{1}{p}\,pf(t) = \frac{1}{p}\,f'(t) = \int^t f'(\tau)\,d\tau = f(t) + c , \tag{21}$$

where c is a constant. To remove the indeterminateness of the constant in (21), let us change the definition of p^{-1} to the definite integral from 0 to t:

$$\frac{1}{p}\,f(t) = \int_0^t f(\tau)\,d\tau . \tag{22}$$

Definition (22) still leads to $p(1/p)f(t) = f(t)$, but now

$$\frac{1}{p}\,pf = If(t) - f(0) , \tag{23}$$

so that $(1/p)p = p(1/p)$ if the function that is operated on vanishes at the origin. Thus if $f(0) = 0$, we can use essentially algebraic methods. However, it would be more advantageous to have definitions such that $p^{-1} \cdot p = p \cdot p^{-1} = I$ for all functions. To do this we must eliminate the $f(0)$ term.

The theory of Laplace transforms is essentially concerned only with functions that are zero for negative values of the argument. To make this explicit, we introduce the Heaviside unit function $H(t)$ which is defined to be equal to one for $t > 0$, equal to zero for $t < 0$, and undefined for $t = 0$. See Figure 6.2:

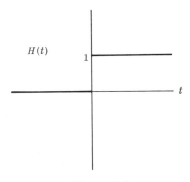

$$H(t)$$

Figure 6.2

Thus a function $f(t)H(t)$ is identically zero for $t < 0$ and is equal to $f(t)$ for $t > 0$. We shall always be careful to write $f(t)H(t)$ for the functions to which we apply the Laplace transform because this device will help us to remove the difficulty that p and p^{-1} do not commute.

Consider $pH(t)$. The derivative $H'(t)$ of $H(t)$ is identically zero for $t < 0$ and $t > 0$ but does not exist in the usual sense for $t = 0$. Were we to put $pH(t) = 0$ for all t, then

$$p^{-1}pH(t) = p^{-1}0 = 0,$$

whereas

$$pp^{-1}H(t) = ptH(t) = H(t), \tag{24}$$

and we would still retain the difficulty that $p^{-1}p \neq pp^{-1}$. However, in view of Chapter 1, we can say that the generalized derivative of $H(t)$ exists at $t = 0$, and write

$$pH(t) = \delta(t), \tag{25}$$

where $\delta(t)$ is the Dirac delta function. For present purposes, $\delta(t)$ is a symbol such that

$$p^{-1}\delta(t) = H(t), \tag{26}$$

and we shall develop additional properties of $\delta(t)$ as we need them. Because of (25) and (26), we have

$$p^{-1}pH(t) = p^{-1}\delta(t) = H(t) \tag{27}$$

and thus

$$p^{-1}pH(t) = pp^{-1}H(t) = H(t). \tag{28}$$

Let us see what additional properties we require of $\delta(t)$ in order for $p^{-1}p[f(t)H(t)]$ to equal $pp^{-1}[f(t)H(t)]$ for an arbitrary differentiable function $f(t)$. We have

$$pp^{-1}[f(t)H(t)] = p\int_0^t f(\tau)d\tau\, H(t) = f(t)H(t). \tag{29}$$

To construct $p^{-1}p[\ ]$, we form

$$p[f(t)H(t)] = f'(t)H(t) + f(t)\delta(t)$$

and then integrate:

$$p^{-1}p[f(t)H(t)] = \int_0^t f'(\tau)\,d\tau\, H(t) + \int_0^t f(\tau)\delta(t)\,dt$$

$$= f(t)H(t) - f(0)H(t) + \int_0^t f(\tau)\delta(\tau)d\tau. \tag{30}$$

Thus if we now assume that

$$f(t)\delta(t) = f(0)\delta(t), \tag{31}$$

then because of (26) we find that

$$\int_0^t f(\tau)\delta(\tau)d\tau = f(0)\int_0^t \delta(\tau)d\tau = f(0)H(t)$$

and consequently (30) reduces to

$$p^{-1}p[f(t)H(t)] = f(t)H(t). \tag{32}$$

Thus by means of $\delta(t)$ defined by (25) and (31), we have $p^{-1}p = pp^{-1} = I$ for $f(t)H(t)$, where $f(t)$ is differentiable.

As examples of this kind of differentiation,

$$p(tH(t)) = H(t) + t\,\delta(t) = H(t) \tag{33}$$

because of (31), and also

$$p^2[tH(t)] = pH(t) = \delta(t). \tag{34}$$

From (34), by division,

$$t\,H(t) = \frac{1}{p^2}\,\delta(t), \tag{35}$$

which we verify by noting that

$$p^{-1}\,\delta(t) = \int_0^t \delta(\tau)d\tau = H(t),$$

and

$$p^{-1}\,H(t) = \int_0^t H(\tau)d\tau = \int_0^t d\tau = tH(t).$$

Another useful result is given by

$$p(e^{at}H(t)) = ae^{at}H(t) + \delta(t) \tag{36}$$

which we "invert" by subtraction

$$(p - a)e^{at} H(t) = \delta(t)$$

and by division

$$e^{at}H(t) = \frac{1}{p - a} \delta(t). \tag{37}$$

If we have a relation

$$f(t)H(t) = F(p)\delta(t), \tag{38}$$

we call $F(p)$ the *Laplace transform* of $f(t)$. The usual definition of the Laplace transform is

$$F(p) = \int_0^\infty e^{-pt} f(t)dt. \tag{39}$$

Later we shall see the connection between the two definitions (38) and (39). Occasionally, instead of (39), the function $F_1(p)$ defined by

$$F_1(p) = p \int_0^\infty e^{-pt} f(t)dt \tag{40}$$

is called the Laplace transform of $f(t)$. The idea behind definition (40) may be seen from the following. We have

$$f(t)H(t) = F(p)\delta(t) = pF(p)H(t) = F_1(p)H(t).$$

We see that $f(t)$ and $pF(p) = F_1(p)$ are analogs in that both act on $H(t)$, and so it seems natural to consider $f(t)$ and $F_1(p)$ together.

As an example, we find the Laplace transform of t^n.

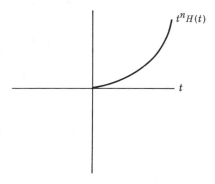

Figure 6.3

We start with the function $t^n H(t)$ of Figure 6.3. Using (25) and (31) we have

$$p t^n H(t) = n t^{n-1} H(t) + t^n \delta(t) = n t^{n-1} H(t) \quad \text{if } n \geq 1,$$

and by successive differentiations, we get

$$p^n t^n H(t) = n! H(t);$$

then by one more differentiation and by use of (25),

$$p^{n+1} t^n H(t) = n! \delta(t),$$

so that

$$t^n H(t) = \frac{n!}{p^{n+1}} \delta(t). \tag{41}$$

Thus the Laplace transformation for t^n is $\dfrac{n!}{p^{n+1}}$.

We shall use the Laplace transform to solve the differential equation

$$u' + u = t$$

with the condition $u(0) = 1$. Consider the function $u(t)H(t)$; then

$$p u(t) H(t) = u'(t) H(t) + \delta(t),$$

where we used (31) to reduce $u(t)\delta(t)$ to $u(0)\delta(t)$, and then used $u(0) = 1$. Similarly,

$$(p + 1) u(t) H(t) = u'(t) H(t) + \delta(t) + u(t) H(t) = t H(t) + \delta(t),$$

by use of the differential equation. By division we get

$$u(t) H(t) = \frac{1}{p + 1} \left[t H(t) + \delta(t) \right] = e^{-t} + \frac{1}{p + 1} t H(t)$$

from (37) for $a = -1$. Now we must interpret $\dfrac{1}{p + 1} t H(t)$. Using (41) for $n = 1$, we have

$$\frac{1}{p + 1} \left[t H(t) \right] = \frac{1}{p + 1} \left[\frac{1}{p^2} \delta(t) \right] = \left[\frac{1}{p + 1} + \frac{1}{p^2} - \frac{1}{p} \right] \delta(t) = \left[e^{-t} + t - 1 \right] H(t)$$

by using (37) and (41); therefore

$$u(t) = \left[2 e^{-t} + t - 1 \right].$$

Another useful Laplace transform is obtained by starting with

$$p (\cos at) H(t) = -a (\sin at) H(t) + \delta(t),$$

where we used $(\cos at)\delta(t) = \delta(t)$. Differentiating again,

$$p^2 (\cos at) H(t) = -a^2 (\cos at) H(t) + p \delta(t),$$

transposing

$$(p^2 + a^2)(\cos at)\, H(t) = p\delta(t),$$

and then dividing, we obtain

$$(\cos at)\, H(t) = \frac{p}{p^2 + a^2}\, \delta(t). \qquad (42)$$

Consequently the operation $\dfrac{p}{p^2 + a^2}$ is the Laplace transform for the cosine. This result may also be obtained by using the integral

$$\int_0^\infty e^{-pt} \cos at\, dt = \frac{p}{p^2 + a^2}.$$

We now illustrate the use of these methods to solve a system of differential equations. Consider

$$u' = u + 2v,$$

$$v' = u + 4v$$

with the initial conditions $u(0) = 1$, $v(0) = 0$. Because of the initial conditions, we have

$$pu(t)H(t) = u'H(t) + \delta(t)$$

and

$$pv(t)H(t) = v'H(t).$$

Multiplying the differential equations by $H(t)$ and adding $\delta(t)$ to the first, we obtain

$$puH(t) = (u + 2v)H(t) + \delta(t),$$

$$pvH(t) = (u + 4v)H(t),$$

or equivalently,

$$(p - 1)uH(t) - 2vH(t) = \delta(t), \qquad uH(t) + (p - 4)vH(t) = 0. \qquad (43)$$

Solving for $uH(t)$ we get

$$[(p - 1)(p - 4) + 2]uH(t) = (p - 4)\delta(t),$$

or

$$(p^2 - 5p + 6)uH(t) = (p - 4)\delta(t);$$

therefore

$$uH(t) = \frac{p - 4}{(p - 2)(p - 3)}\, \delta(t) = \left[\frac{2}{p - 2} - \frac{1}{p - 3}\right]\delta(t) = (2e^{2t} - e^{3t})H(t)$$

from (37). To find v, we use the first equation of (43):

$$2vH(t) = (p - 1)uH(t) - \delta(t) = \left[4e^{2t} - 3e^{3t} - 2e^{2t} + e^{3t}\right]H(t)$$

$$+ 2e^{2t}\delta(t) - e^{3t}\delta(t) - \delta(t) = 2(e^{2t} - e^{3t})H(t).$$

Thus the solution is

$$u = 2e^{2t} - e^{3t}, \qquad v = e^{2t} - e^{3t}.$$

These examples show that in order to solve a differential equation or a system of differential equations, it is necessary to interpret the function of p acting on the delta function. If $R(p)$ is a rational function (a ratio of polynomials) of p acting on the delta function, then we interpret

$$u = R(p)\delta(t)$$

by expanding $R(p)$ in partial fractions. If $R(p)$ gives only simple factors, we work with

$$u = \left[\sum \frac{A_K}{p - \alpha_K}\right]\delta(t)$$

$$= \sum A_K e^{\alpha_K t}H(t).$$

To evaluate repeated factors, consider $\dfrac{1}{(p - \alpha)^2}\delta(t)$. Using $e^{\alpha t}\delta(t) = \delta(t)$, we write

$$\frac{1}{(p - \alpha)^2}\delta(t) = \frac{1}{(p - \alpha)^2}e^{\alpha t}\delta(t).$$

But, by the shift rule (4), we have

$$F(p)e^{\alpha t}u(t) = e^{\alpha t}F(p + \alpha)u(t)$$

so that

$$\frac{1}{(p - \alpha)^2}e^{\alpha t}\delta(t) = e^{\alpha t}\frac{1}{p^2}\delta(t) = e^{\alpha t}tH(t),$$

where the final form followed from (41). Thus

$$\frac{1}{(p - \alpha)^2}\delta(t) = te^{\alpha t}H(t). \tag{45}$$

Convolution Theorem. Suppose we know a pair of transforms

$$F(p)\delta(t) = f(t)H(t), \qquad G(p)\delta(t) = g(t)H(t), \tag{46}$$

and would like to find the inverse Laplace transform of the product $F(p)G(p)$, that is, the function $u(t)$ such that

$$F(p)G(p)\delta(t) = u(t)H(t). \tag{47}$$

Using $f(\tau)\delta(t - \tau) = f(t)\delta(t - \tau)$, and

$$\int_{-\infty}^{\infty} f(\tau) H(t)\delta(t - \tau)\, d\tau = f(t)H(t) \int_{0}^{\infty} \delta(t - \tau)\, d\tau = f(t)H(t)\,,$$

we have

$$G(p)[F(p)\delta(t)] = G(p)[f(t)H(t)] = G(p)\int\delta(t - \tau)f(\tau)H(\tau)\, d\tau$$

$$= \int [G(p)\,\delta(t - \tau)]\,f(\tau)H(\tau)\, d\tau$$

$$= \int_{-\infty}^{\infty} [g(t - \tau)H(t - \tau)]\,f(\tau)H(\tau)\, d\tau$$

$$= \int_{0}^{t} g(t - \tau)f(\tau)\, d\tau\, H(t)\,.$$

Thus

$$F(p)\,G(p)\delta(t) = \int_{0}^{t} g(t - \tau)f(\tau)\, d\tau\, H(t) = u(t)H(t)\,. \tag{48}$$

The function $u(t)$ is called the convolution of $f(t)$ and $g(t)$. We have thus proved that the Laplace transform of the convolution is the product of the Laplace transforms. This theorem is valid because differentiation commutes with translation. To see this, put T for the translation operator so that

$$Tu(t) = u(t + \tau)\,;$$

then

$$\frac{d}{dt}\, Tu = T\frac{d}{dt}\, u(t)\,.$$

A few other properties of the Laplace transform $f(t)H(t) = F(p)\delta(t)$ will now be obtained. Let us consider $\lim_{t\to\infty} f(t)$ and $\lim_{t\to 0} f(t)$ for rational $F(p)$. We write $F(p) = P(p)/Q(p)$, and assume that if α_k is such that $Q(\alpha_k) = 0$, then Re $\alpha_k \le 0$. [If Re $\alpha_k > 0$, then $\lim_{t\to\infty} f(t) = \infty$.] For all α_k, Re $\alpha_k < 0$, we have $\lim_{t\to\infty} f(t) = 0$, but if Re $\alpha_k = 0$, then the result is nonvanishing. Suppose $\alpha_0 = 0$ is a simple root of $Q(\alpha_k) = 0$, and for all other roots Re $\alpha_k < 0$; then write

$$F(p) = \frac{A}{p} + \sum \frac{b_k}{p - \alpha_k} + \cdots\,,$$

$$f(t)H(t) = A H(t) + \sum b_k\, e^{\alpha_k t}\, H(t)$$

and since

$$pF(p) = A + p\sum\,,$$

we see that

$$\lim_{t\to\infty} f(t) = \lim_{p\to 0} p\, F(p)\,, \tag{49}$$

in both cases the result being A. If zero is a multiple root of $Q(p)$, then $\lim_{t \to \infty} f(t)$ will not exist.

For $p \to \infty$, it follows from

$$pf(t)H(t) = f'(t)H(t) + f(0)\delta(t) = pF(p)\delta(t) = \left(A + \sum_{1}^{\infty} \frac{a_k}{p^k} \right)\delta(t)$$

that

$$f(0) = \lim_{t \to 0} f(t) = \lim_{p \to \infty} pF(p). \tag{50}$$

As an illustration of this result, consider the cosine transform given in (42):

$$f(t)H(t) = (\cos at)\, H(t) = \frac{p}{p^2 + a^2}\, \delta(t) = F(p)\delta(t),$$

for which $pF(p) = \dfrac{p^2}{p^2 + a^2}$. We see that

$$\lim_{t \to 0} \cos at = \lim_{p \to \infty} \frac{p^2}{p^2 + a^2} = 1.$$

So far, we have discussed only rational functions of p. We shortly consider irrational functions of p, such as

$$p^{-1/2}\delta(t) = f(t)H(t). \tag{51}$$

Before turning to this, we consider the equation

$$tF(p)\delta(t) = tf(t)H(t).$$

We would like to move t to the other side of the operator $F(p)$, and therefore seek something similar to the shift rule. Since

$$p(tu(t)) = tpu(t) + u(t),$$

we get

$$pt \doteqdot tp + I \tag{52}$$

where the symbol $\doteqdot$ stresses that this is a relation between operators. Similarly

$$p^2 t \doteqdot p(pt) \doteqdot p(tp + I)$$

$$\doteqdot p(tp) + p$$

$$\doteqdot (tp + I)p + p$$

$$\doteqdot tp^2 + 2p,$$

and we have the general relation

$$p^n t \doteqdot tp^n + np^{n-1}, \qquad tp^n \doteqdot p^n t - np^{n-1}. \tag{53}$$

Thus for polynomials, it follows from (53) that

$$tF(p) \doteq F(p)t - F'(p),\tag{54}$$

and consequently

$$tF(p)\delta(t) = F(p)t\delta(t) - F'(p)\delta(t) = 0 - F'(p)\delta(t).\tag{55}$$

For irrational functions, for example, $F(p) = p^{-1/2}$, we find that

$$tf(t)H(t) = tp^{-1/2}\,\delta(t) = p^{-1/2}\,t\delta(t) + \frac{1}{2}\,p^{-3/2}\delta(t) = +\frac{1}{2}\,p^{-3/2}\,\delta(t).$$

But

$$p[tf(t)H(t)] = +\frac{1}{2}p^{-1/2}\delta(t) = +\frac{1}{2}f(t)H(t);$$

consequently

$$(tf)'H(t) = +\frac{1}{2}f(t)H(t),$$

or

$$tf' + f = +\frac{1}{2}f,$$

or

$$tf' = -\frac{1}{2}f.$$

The solution of this equation is $f = Ct^{-1/2}$. Thus

$$p^{-1/2}\,\delta(t) = Ct^{-1/2}H(t)$$

We still need to find the value of the constant C. Using the convolution theorem (48), we have

$$p^{-1}\delta(t) = p^{-1/2} \cdot p^{-1/2}\,\delta(t) = \int_0^t C^2(t-\tau)^{-1/2}\tau^{-1/2}\,d\tau\,H(t) = \pi C^2 H(t);$$

but $p^{-1}\delta(t) = H(t)$, so we have found

$$\pi C^2 = 1, \qquad C = \frac{1}{\sqrt{\pi}},$$

and consequently

$$p^{-1/2}\,\delta(t) = \frac{1}{\sqrt{\pi t}}\,H(t).\tag{56}$$

Equation (56) can be used to obtain the fractional integral of a function. We write the Laplace transform as $u(t)H(t) = U(p)\delta(t)$, and obtain

$$p^{-1/2}\,u(t)H(t) = p^{-1/2}\,U(p)\delta(t) = \frac{1}{\sqrt{\pi}}\int_0^t \frac{1}{\sqrt{t-\tau}}\,u(\tau)\,d\tau.\tag{57}$$

The right-hand side of this equation is called the one-half integral of $u(t)$.

It arises, for example, in *Abel's integral equation*, for which we are given $f(t)$ and seek $u(\tau)$ such that

$$f(t) = \frac{1}{\sqrt{\pi}} \int_0^t \frac{u(\tau)\, d\tau}{\sqrt{t-\tau}} .$$

We may write this equation as

$$f(t)H(t) = p^{-1/2} u(t)H(t) .$$

Then

$$u(t)H(t) = p^{1/2} f(t)H(t) = pp^{-1/2} f(t)H(t)$$

$$= p\, \frac{1}{\sqrt{\pi}} \int_0^t \frac{f(\tau)\, d\tau}{\sqrt{t-\tau}} = \frac{d}{dt}\, \frac{1}{\sqrt{\pi}} \int_0^t \frac{f(\tau)\, d\tau}{\sqrt{t-\tau}} .$$

6.4 SYMBOLIC METHODS AND GENERATING FUNCTIONS

A sequence of numbers

$$a_0 = 1,\ a_1,\ a_2,\ \ldots \tag{58}$$

or a sequence of functions

$$f_0(x) = 1,\ f_1(x),\ f_2(x),\ \ldots \tag{59}$$

can be treated more conveniently by means of symbolic generating functions we associate with

$$\sum_0^\infty a_n t^n, \quad \sum a_n t^n/n!, \quad \sum f_n(x)t^n, \quad \text{or} \quad \sum f_n(x)t^n/n!$$

The convenience is due to the fact that it is easier to handle one symbol than an infinite number of terms in a sequence. In addition, we shall show how to use various symbolic methods to further simplify the manipulations. For example, the sequence (58) will be represented by the symbol A and the sequence (59) by the symbol F. We shall define rules for operating with these symbols and then use these rules to obtain properties of the sequences.

If A and B are symbols representing the sequences

$$1,\quad a_1,\quad a_2,\ \ldots$$

and

$$1,\quad b_1,\quad b_2,\ \ldots ,$$

respectively, and if

$$f(t,s) = \sum\sum_0^\infty \alpha_{jk} t^j s^k ,$$

where the series converges in some neighborhood of $t = 0$, $s = 0$, then we shall use $f(A, B)$ to denote the formal series

$$\sum\sum \alpha_{jk} A^j B^k . \tag{60}$$

To each symbol $f(A,B)$ we associate the number obtained from (60) when A^j is replaced by a_j and B^k is replaced by b_k. We represent this association by the symbol $\doteq$, and thus we write

$$f(A,B) \doteq \sum\sum \alpha_{jk} a_j b_k , \tag{61}$$

where we assume the series on the right-hand side is convergent. We write also

$$\sum\sum \alpha_{jk} a_j b_k \doteq f(A,B) .$$

We say two symbols $f(A,B)$ and $g(A,B)$ are *equal* if they are associated with equal numbers. Thus, if

$$g(A,B) = \sum\sum \beta_{jk} A^j B^k$$

then

$$f(A,B) = g(A,B)$$

if and only if,

$$\sum\sum \alpha_{jk} a_j b_k = \sum\sum \beta_{jk} a_j b_k .$$

Because of these definitions, the symbols A,B, etc., can be considered as elements of an algebra.

We introduce now the concept of identical symbols. Two symbols A and B are said to be *identical*, written $A \equiv B$ if and only if

$$\varphi(A) = \varphi(B) \tag{62}$$

for any function $\varphi(t)$. Note that (62) is equivalent to

$$a_n = b_n$$

for $n = 0, 1, 2, \ldots$. Finally, we define a new kind of sum of two symbols. The *fold* of two symbols A and B, written $A \oplus B$, is the symbol C such that

$$e^{tC} = e^{tA} e^{tB} . \tag{63}$$

By comparing corresponding powers of t on both sides we see that

$$C^n = (A + B)^n, \quad n = 0, 1, 2, \ldots \tag{64}$$

or

$$c_n = \sum_0^n \binom{n}{k} a_k b_{n-k} \qquad n = 0, 1, 2, \ldots . \tag{65}$$

We assume that A is not identical with B, so that we write

$$C \equiv A \oplus B \equiv A + B,$$

but note that

$$A \oplus A \neq A + A \equiv 2A .$$

We illustrate these ideas by considering the extended Bernoulli numbers defined by the coefficients in the expansion

$$\frac{t}{e^t - 1} = \sum_0^\infty B_k \frac{t^k}{k!} ,$$

that is, $B_0 = 1$, $B_1 = -\frac{1}{2}$, $B_2 = \frac{1}{6}$, $\ldots$, $B_{2n+1} = 0$. Let us use the symbol B for these numbers so that $B^k = B_k$, the kth Bernoulli number. Then, since

$$\frac{t}{e^t - 1} = \sum_0^\infty B_k \frac{t^k}{k!} \doteq e^{Bt} , \tag{66}$$

we say the function $t(e^t - 1)^{-1}$ is the *exponential generating function* for the Bernoulli numbers. Let us use I as a symbol to denote the sequence of numbers all of which are unity; then

$$e^{It} \doteq e^t . \tag{67}$$

Using this in (66), we get

$$t = e^{Bt}(e^{It} - 1) = e^{Bt} e^{It} - e^{Bt} = e^{(B \oplus I)t} - e^{Bt} .$$

Comparing corresponding powers of t, we find that

$$(B \oplus I)^n - B^n = (B + I)^n - B^n = 0, \qquad n > 1 ; \tag{68}$$

and for $n = 1$

$$B + I - B \doteq 1 .$$

The formula (68) enables us to find the Bernoulli numbers successively. We have for $n = 2, 3, \ldots$

$$O = B^2 + 2BI + I^2 - B^2 \doteq B_2 + 2B_1 + 1 - B_2 = 0, \text{ or } B_1 = -1/2;$$

$$O = B^3 + 3B^2 I + 3BI^2 + I^3 - B^3 \doteq B_3 + 3B_2 + 3B_1 + 1 - B_3$$

$$= 0, \text{ or } B_2 = 1/6 ,$$

and so on.

Consider the Bernoulli polynomials $B_k(x)$ defined by

$$\frac{te^{xt}}{e^t - 1} = \sum_0^\infty B_k(x)\, \frac{t^k}{k!} \doteq e^{tB(x)}, \tag{69}$$

where we have used $B(x)$ as a symbol for the sequence of Bernoulli polynomials, $1, B_1(x), B_2(x), \ldots$; we want an expression for $B_k(x)$ in terms of
x^n and B_n. Comparing (69) with (66) we see that

$$e^{tB(x)} = e^{xt}\, e^{Bt} = e^{t(xI + B)};$$

we indicate the equivalence of the symbols in the exponents by

$$B(x) \equiv xI + B. \tag{70}$$

Taking successive powers of (70), and converting, we get

$$B_1(x) = x + B_1,$$

$$B_2(x) = x^2 + 2xB_1 + B_2,$$

and so on.

We now derive a formula for the first difference of a Bernoulli
polynomial. From

$$B(x) \equiv xI + B \equiv (x - 1)I + (I + B), \tag{71}$$

we get

$$B^n(x) \equiv [(x - 1)I + (I + B)]^n$$

$$= (x - 1)^n + n(x - 1)^{n-1}(I + B) + \binom{n}{2}(x - 1)^{n-2}(I + B)^2 + \cdots,$$

which by (68) for $n > 1$ reduces to

$$B^n(x) = (x - 1)^{n-1} + n(x - 1)^{n-1}(1 + B) + \binom{n}{2}(x - 1)^{n-2} B^2 + \cdots$$

$$\equiv [(x - 1)I + B]^n + n(x - 1)^{n-1},$$

and, finally, by (70) to

$$B^n(x) \equiv B^n(x - 1) + n(x - 1)^{n-1}$$

Thus the Bernoulli polynomials satisfy

$$B_n(x) - B_n(x - 1) = n(x - 1)^{n-1}. \tag{72}$$

Using this equation for $x = 1, 2, 3, \ldots, k, \ldots, N$, and then adding the
resulting set of equations, we obtain the summation formula

$$B_n(N) - B_n(0) = n \sum_1^{N-1} k^{n-1}, \tag{73}$$

where $B_n(0)$ is the Bernoulli number B_n .

Consider the sequence of numbers

$$f_0, f_1, f_2, \cdots$$

and let F be the symbol for this sequence. The exponential generating function is

$$f(t) = \sum_0^\infty \frac{f_k t^k}{k!} \doteq e^F t . \tag{74}$$

Consider the function

$$e^t f(xt) = g(x,t) = \sum_0^\infty g_k(x) \frac{t^k}{k!} \doteq e^{tG(x)} \tag{75}$$

so that $G(x)$ is the symbol for the sequence of polynomials $g_n(x)$. We have

$$e^{tG(x)} = e^{It}\, e^{xtF} ;$$

therefore

$$G(x) \equiv I \oplus xF . \tag{76}$$

From (64) and (65) this implies that

$$g_n(x) = \sum_0^n \binom{n}{k} x^k f_k .$$

We now want to find the recurrence relationship between g_n and the derivatives of g_n. We use $G'(x)$ as a symbol for the sequence $g_n'(x)$, and write

$$e^{tG'(x)} \doteq \sum \frac{t^k}{k!}\, g_k'(x)$$

$$\doteq \frac{d}{dx}\, e^{tG(x)} = tFe^{t(I \oplus xF)} = tFe^{tG(x)} , \tag{77}$$

and eliminate F to obtain a relationship entirely in terms of G. From (76)

$$G(x) - I \equiv xF ;$$

then

$$F \equiv \frac{G(x) - I}{x}$$

Substituting in (77) we get

$$e^{tG'(x)} = \frac{t[G(x) - I]}{x}\, e^{tG(x)}$$

or

$$xe^{tG'(x)} = tG(x)e^{tG(x)} - te^{tG(x)} = t\frac{d}{dt}e^{tG(x)} - te^{tG(x)}.$$

Comparing corresponding powers of t in the associated expansions, we obtain

$$xg_n'(x) = ng_n(x) - ng_{n-1}(x).\tag{78}$$

Similarly we find a formula for $g_n(xy)$ in terms of $g_n(x)$. We have

$$G(xy) \equiv I + xyF \equiv I - yI + yI + xyF \equiv (1 - y)I + y\,(I + xF)$$
$$\equiv (1 - y)I + yG(x).\tag{79}$$

Taking the nth power of both sides and converting, we get

$$g_n(xy) = \sum \binom{n}{k} y^k g_k(x)(1 - y)^{n-k}.\tag{80}$$

We shall now find the exponential generating function for the Legendre polynomials. We write

$$\sum_0^\infty P_n(\mu)\frac{t^n}{n!} = P(t,\mu).\tag{81}$$

It is a well-known property of Laplace's equation that $\nabla^2[r^n P_n(\cos\theta)] = 0$, for all values of n; here ∇^2 is Laplace's operator, r is the distance from the origin, and θ is the polar angle. If we identify r with t and $\cos\theta$ with μ, then

$$\nabla^2 P(t,\mu) = 0.\tag{82}$$

To solve this partial differential equation, we need some boundary conditions. Since

$$P_n(1) = 1, \qquad P_n(-1) = (-1)^n$$

we see that for $\mu = 1$ and $\mu = -1$, we have from the definition of $P(t,\mu)$ that

$$P(t,1) = e^t = e^r = e^z, \qquad P(t,-1) = e^{-t} = e^{-r} = e^{-z}.\tag{83}$$

From (82) and (83), $P(t,u)$ is a harmonic function whose values are known on the z axis; this suggests introducing cylindrical coordinates ρ, φ, z. A well-known solution of Laplace's equation which behaves like an exponential on the z axis is $e^{\alpha z} J_0(\alpha\rho)$. If we take $\alpha = 1$, the function $e^z J_0(\rho)$ will equal e^z for $\rho = 0$, that is, on the z axis. We see then that $P(t,\mu) = e^z J_0(\rho)$. Since $z = r\cos\theta = t\mu$ and $\rho = r\sin\theta = t\sqrt{1 - \mu^2}$, we have, finally

$$P(t,\mu) = e^{t\mu} J_0(t\sqrt{1 - \mu^2}) = \sum \frac{P_n(\mu)t^n}{n!}.\tag{84}$$

The usual generating function for the Legendre polynomials is

$$\sum_0^\infty P_n(\mu)\, t^n = (1 - 2t\mu + t^2)^{-1/2} \tag{85}$$

One can easily arrive at this expression by working with symbolic generating functions. From (84) we have

$$e^{t\mu}\, J_0\left(t\sqrt{1 - \mu^2}\right) = \sum_{n=0}^\infty \frac{P_n(\mu)\, t^n}{n!} \doteq e^{tP(\mu)}, \tag{86}$$

which is our standard symbolic generator. Similarly from $\sum p^n t^n = \dfrac{1}{1 - tp}$, we introduce the symbolic form

$$\sum_0^\infty P_n(\mu)\, t^n \doteq \frac{1}{1 - tP(\mu)}. \tag{87}$$

The symbolic generator of (87) can be represented as the integral

$$\int_0^\infty e^{-s[1-tP(\mu)]}\, ds = \int_0^\infty e^{stP(\mu)}\, e^{-s}\, ds; \tag{88}$$

using (86), we replace the symbolic exponential $e^{stP(\mu)}$ to obtain

$$\frac{1}{1 - tP(\mu)} \doteq \int_0^\infty e^{-s(1-t\mu)}\, J_0\left(st\sqrt{1 - \mu^2}\right) ds,$$

and since

$$\int_0^\infty e^{-\alpha s}\, J_0(\beta s)\, ds = (\alpha^2 + \beta^2)^{-1/2},$$

we find that

$$\frac{1}{1 - tP\mu} \doteq \left[(1 - t\mu)^2 + t^2(1 - \mu^2)\right]^{-1/2} = (1 - 2\mu t + t^2)^{-1/2}. \tag{89}$$

Next we consider the Hermite polynomials defined by the generating function

$$e^{2xt - t^2} = \sum_0^\infty H_n(x)\, \frac{t^n}{n!} \doteq e^{tH(x)}. \tag{90}$$

Using symbolic methods, we write

$$e^{-t^2} = \sum \frac{(-t^2)^n}{n!} = \sum a_k \frac{t^k}{k!} \doteq e^{tA} \tag{91}$$

where $a_k = 0$ for k odd, and $a_k = (-1)^n (2n)!/n!$ for $k = 2n$.

Now (90) may be written as

$$e^{tH(x)} \doteq e^{2xt}\, e^{-t^2} \doteq e^{2xtI}\, e^{tA};$$

therefore

$$H(x) \equiv 2xI \oplus A.$$

Taking the nth power of both sides, and converting, we get

$$H_n(x) = \sum \binom{n}{k} (2x)^k a_{n-k} . \qquad (92)$$

Consider

$$H(x + y) \equiv 2(x + y)I \oplus A \equiv 2yI \oplus 2xI \oplus A \equiv H(x) \oplus 2yI.$$

Taking the nth power of this formula, and converting, we have

$$H_n(x + y) = \sum_0^n \binom{n}{k} H_k(x)(2y)^{n-k} . \qquad (93)$$

We can find the Hermite polynomial of a Legendre polynomial as follows:

$$H(x + P(x)) = 2xI \odot 2P(x) \oplus A$$

$$= H(x) \oplus 2P(x).$$

Then, again by taking nth powers, and converting, we find

$$H_n(x + P(x)) = \sum \binom{n}{k} H_k(x) \, 2^{n-k} P_{n-k}(x) . \qquad (94)$$

As a final illustration of these methods, consider

$$H(x) \oplus H(y) \equiv 2xI \oplus A \oplus 2yI \oplus A \equiv 2(x + y)I \oplus A \oplus A. \qquad (95)$$

Since $A \oplus A \neq 2A$, the problem of evaluating $A + A$ arises. By definition, and by (91),

$$e^{tA} e^{tA} = e^{t(A \oplus A)} \doteq e^{-t^2} e^{-t^2} = e^{-2t^2} \doteq e^{\sqrt{2} tA} ;$$

therefore

$$A \oplus A \equiv \sqrt{2} A .$$

Using this in (95), we get

$$H(x) \oplus H(y) \equiv 2(x + y)I \oplus \sqrt{2}A \equiv \sqrt{2}\left[\frac{2(x + y)}{\sqrt{2}} \oplus A \right]$$

$$\equiv \sqrt{2} \, H \left(\frac{x + y}{\sqrt{2}} \right).$$

Taking nth powers, and converting, we get the addition theorem for Hermite polynomials:

$$H_n \left(\frac{x + y}{\sqrt{2}} \right) = \frac{1}{2^{n/2}} \sum_{k=0}^n \binom{n}{k} H_k(x) H_{n-k}(y) . \qquad (96)$$

We shall conclude the discussion of Hermite polynomials by obtaining their differential equation from (90). By differentiation we get

$$\sum H_n' \frac{t^n}{n!} = 2t \, e^{2xt-t^2},$$

and

$$\sum H_n'' \frac{t^n}{n!} = 4t^2 e^{2xt-t^2}.$$

Combining these two results, we have

$$\sum (H_n'' - 2xH_n') \frac{t^n}{n!} = (4t^2 - 4xt) \, e^{2xt-t^2}$$

$$= -2t \, \frac{\partial}{\partial t} \, e^{2xt-t^2}$$

$$= \sum (-2n)H_n(x) \frac{t^n}{n!}.$$

Comparing coefficients of t^n on both sides, we obtain

$$H_n'' - 2xH_n' = -2n H_n(x).$$

6.5 NONCOMMUTATIVE OPERATORS

In this section we shall discuss some applications of the theory of non-commutative operators. We begin with the Hermite polynomials. Consider the solutions of

$$\frac{d^2}{dx^2} u + (\lambda - x^2)u = 0 \tag{97}$$

which are bounded for $-\infty < x < \infty$. We may rewrite this relation as the operational equation

$$Lu = \lambda u, \quad L = x^2 - \frac{d^2}{dx^2}. \tag{98}$$

We attempt to write L as the product of two factors $[x + (d/dx)]$ $[x - d/dx)]$, but

$$\left(x + \frac{d}{dx}\right)\left(x - \frac{d}{dx}\right)u = \left(x + \frac{d}{dx}\right)(xu - u')$$

$$= x^2 u - xu' + xu' + u - u''$$

$$= Lu + u.$$

The factorization did not work because the operators $[x + (d/dx)]$ and $[x - (d/dx)]$ do not commute; the cross-terms differ,

$$\left(\frac{d}{dx} x\right) u \neq \left(x \frac{d}{dx}\right) u,$$

because

$$\left(\frac{d}{dx} x\right) u = \left(x \frac{d}{dx}\right) u + u.$$

Despite the failure of the factorization, we introduce the operators A and A^*, where

$$A = x + \frac{d}{dx},$$

$$A^* = x - \frac{d}{dx}.$$

(99)

These operators do not commute. We have

$$A A^* = \left(x + \frac{d}{dx}\right) \left(x - \frac{d}{dx}\right) = L + 1,$$

(100)

and

$$A^* A = \left(x - \frac{d}{dx}\right) \left(x + \frac{d}{dx}\right) = L - 1,$$

(101)

thus

$$A A^* - A^* A = 2.$$

(102)

If A and A^* did commute, the above difference would be zero. The difference is called the *commutator* of A and A^*. We shall use the notation

$$[A, A^*] = A A^* - A^* A$$

(103)

This bracket operation has the following properties:

(a) $[A, A] = 0,$

(b) $[A, (\text{constant})] = 0,$

(c) $[B, A] = - [A, B],$

(d) $[\alpha_1 A + \beta_1 B, \alpha_2 A + \beta_2 B] = \begin{vmatrix} \alpha_1 & \beta_1 \\ \alpha_2 & \beta_2 \end{vmatrix} [A, B],$

(e) $[[A, B], C] + [[B, C], A] + [[C, A], B] = 0$

Property (e) is known as *Jacobi's identity*.

Suppose $\mathcal{L}$ is a collection of elements $A, B, C, \ldots$ such that linear combinations $\alpha A + \beta B + \gamma C + \ldots$ and bracket operations $[A, B]$ are defined. Suppose $\mathcal{L}$ contains all linear combinations such as $\alpha A + \beta B + \gamma C + \ldots$ and suppose that any bracket is a linear combination of elements of $\mathcal{L}$; that is, suppose

$$[A, B] = \alpha A + \beta B + \gamma C + \ldots . \tag{104}$$

Then $\mathcal{L}$ is called a *Lie algebra* generated by the elements $A, B, \ldots$.

We show that the operators $A, A*$ and L generate a Lie algebra. We have from (102) and (103),

$$[A, A*] = 2 . \tag{105}$$

Using L of (100), we find that

$$[L, A] = [AA* - 1, A] = [AA*, A]$$

$$= AA*A - AAA*$$

$$= A[A*A - AA*]$$

$$= A[A*, A] = -2A ,$$

that is,

$$[L, A] = -2A . \tag{106}$$

Similarly,

$$[L, A*] = 2A* . \tag{107}$$

Then, since all commutators of $A, A*$, and L can be expressed in terms of $A, A*$ we have a Lie algebra.

Suppose we have one eignefunction of L , i.e., a solution of $Lu = \lambda u$. If we have just one eigenfunction, we can find others by using the operators A and $A*$. We find by definition of $[A, B]$ and by use of (106) that

$$L(Au) = ALu + [L,A]u = ALu - 2Au$$

$$= (\lambda - 2)Au . \tag{108}$$

This shows that if $Au \neq 0$, it is an eigenfunction of L corresponding to the eigenvalue $\lambda - 2$. Similarly, by use of (107),

$$L(A*u) = A*Lu + [L,A]u$$

$$= \lambda A*u + 2A*u \tag{109}$$

$$= (\lambda + 2) A*u .$$

Thus, if $A*u \neq 0$, it is an eigenfunction of L corresponding to the eigenvalue $\lambda + 2$. By repeated application of the operators A (or $A*$) we get a series of eigenfunctions for smaller and smaller (or larger and larger) values of λ, unless at some stage we find an eigenfunction v such that either $Av = 0$ (or $A*v = 0$).

But do there exist any functions such that Av or A^*v will equal zero? To find such a function we must find a solution of $Av = xv + v' = 0$ or $A^*v = xv - v' = 0$. The solution of $Av = 0$ is $v = e^{-x^2/2}$. The solution of $A^*v = 0$ is $v = e^{x^2/2}$, but this function is not bounded for $-\infty < x < \infty$ and cannot be an eigenfunction.

Thus we have found one eigenfunction $(u_0 = e^{-x^2/2})$ such that $Au_0 = 0$, and by (101),

$$Le^{-x^2/2} = A^* A e^{-x^2/2} + e^{-x^2/2}$$

or, since $Ae^{-x^2/2} = Au_0 = 0$,

$$Lu_0 = u_0, \qquad \lambda_0 = 1. \tag{110}$$

We now write

$$A^* u_0 = u_1, \qquad A^* u_1 = (A^*)^2 u_0 = u_2, \quad (A^*)^n u_0 = u_n. \tag{111}$$

Then from (110) and (111), by repeated use of (109), we get $Lu_1 = (\lambda_0 + 2)u_1 = 3u_1 = \lambda_1 u_1$, $Lu_2 = (\lambda_1 + 2)u_2 = 5u_2$, etc., so that

$$Lu_n = (1 + 2n)u_n. \tag{112}$$

We now have an eigenfunction corresponding to every odd integer.

Consider the expression

$$A[f(x)e^{-x^2/2}] = x[f(x)e^{-x^2/2}] + f'e^{-x^2/2} - x[f(x)e^{-x^2/2}]$$
$$= f'e^{-x^2/2}, \tag{113}$$

and also

$$A^*[f(x)e^{x^2/2}] = xfe^{x^2/2} - f'e^{x^2/2} - xfe^{x^2/2}$$
$$= -f'e^{x^2/2}. \tag{114}$$

Using these results we shall find a simpler expression for u_n. We have using (114),

$$u_n = (A^*)^n u_0 = (A^*)^{n-1} A^* e^{-x^2/2}$$
$$= (A^*)^{n-1} A^* (e^{-x^2} e^{x^2/2})$$
$$= -(A^*)^{n-1} \left[\frac{d}{dx}(e^{-x^2}) \right] e^{x^2/2}.$$

By repeated use of (114) we obtain

$$u_n = (-1)^n \left[\frac{d^n}{dx^n} e^{-x^2} \right] e^{x^2/2}$$

$$= (-1)^n \left[\frac{d^n}{dx^n} e^{-x^2} \right] e^{x^2} e^{-x^2/2}$$

$$= H_n(x) e^{-x^2/2}$$

where

$$H_n(x) = (-1)^n e^{x^2} \frac{d^n e^{-x^2}}{dx^n} \tag{115}$$

It is easy to see that $H_n(x)$ is a polynomial of the nth degree, the nth Hermite polynomial.

We have obtained the Hermite functions u_n by starting with u_0 and using the operator A^* to increase the eigenvalue. What happens if we use A to decrease the eigenvalue? Since by (108) we have $L(Au_n) = (\lambda_n - 2)Au_n = (2n - 1)Au_n$, we see that Au_n is an eigenfunction of L corresponding to the eigenvalue $2n - 1$; we must therefore have

$$Au_n = \alpha u_{n-1} \tag{116}$$

where α is some constant. But

$$A^* Au_n = \alpha A^* u_{n-1} = \alpha u_n . \tag{117}$$

Using (101), $AA^* = L - 1$, and the fact that

$$Lu_n - u_n = 2nu_n , \tag{118}$$

we see from (117) and (118) that $\alpha = 2n$, and consequently (116) becomes

$$Au_n = 2nu_{n-1} . \tag{119}$$

Now using (113) with $f = H_n(x)$, and applying (119), we get

$$A \left[H_n(x) e^{-x^2/2} \right] = H_n'(x) e^{-x^2/2}$$

$$= 2n H_{n-1}(x) e^{-x^2/2} ;$$

consequently,

$$H_n'(x) = 2n H_{n-1}(x) . \tag{120}$$

The eigenfunctions u_n are mutually orthogonal over the interval $(-\infty, \infty)$. Before proving this, we note that

$$A(A^*)^m = (A^*)^m A + [A, (A^*)^m] \tag{121}$$

by definition of $[A,B]$, and that

$$[A, (A^*)^m] = 2m(A^*)^{m-1} \tag{122}$$

Equation (122) can be easily proved by mathematical induction. In fact, we can show that

$$[A, \varphi(A^*)] = 2 \frac{\partial \varphi(A^*)}{\partial A^*}$$

where $\varphi(A^*)$ is an analytic function of A^*.

Consider the integral

$$I_{nm} = \int_{-\infty}^{\infty} u_n(x) u_m(x) \, dx = \int_{-\infty}^{\infty} (A^*)^n u_0 \cdot (A^*)^m u_0 \, dx, \quad 1 \leq n \leq m,$$

and the result obtained on integration by parts,

$$I_{nm} = \int_{-\infty}^{\infty} (A^*)^{n-1} u_0 \cdot A(A^*)^m u_0 \, dx.$$

From (121) and (122), and the fact that $Au_0 = 0$, we find that

$$I_{nm} = 2m \int_{-\infty}^{\infty} (A^*)^{n-1} u_0 \cdot (A^*)^{m-1} u_0 \, dx = 2m I_{n-1, m-1}. \tag{123}$$

We repeat this process until we get to the integral

$$I_{0,p} = \int_{-\infty}^{\infty} u_0 \cdot (A^*)^p u_0 \, dx$$

where $p = m - n$. If $p = 0$, it is well-known that $I_{0,0} = \sqrt{\pi}$, and we obtain from (123) the result

$$I_{n,n} = \int_{-\infty}^{\infty} u_n^2 \, dx = 2^n n! \sqrt{\pi}. \tag{124}$$

If $p > 0$, then integration by parts shows that

$$I_{0,p} = \int_{-\infty}^{\infty} u_0 (A^*)^p u_0 \, dx = \int_{-\infty}^{\infty} A u_0 \cdot (A^*)^{p-1} u_0 \, dx = 0$$

because of the definition of u_0; therefore

$$I_{n,m} = \int_{-\infty}^{\infty} u_n u_m \, dx = 0, \tag{125}$$

if $n \neq m$. Thus we have shown that the eigenfunctions are mutually orthogonal.

Lie Algebras and Groups. We shall now show the correlation between Lie algebras and groups. We shall discuss the set of all rotations of space about a fixed point called the origin. Note that any succession of rotations can be expressed as a single rotation. The identity rotation (I) is defined as no rotation from the original position.

Consider the set of rotations

$$I, R_1, R_2, \ \ldots \ . \tag{126}$$

We write the rotation R_3 obtained by performing the rotations R_2 and R_1 in that order as $R_3 = R_1 R_2$, and we call R_3 the product of R_1 and R_2. This product obeys the associative law as can be seen by considering the products $(R_1 R_2) R_4$ and $R_1 (R_2 R_4)$. If we denote by R^{-1} the rotation which reverses the effect of R, then we find

$$IR = RI = R,$$
$$R R^{-1} = R^{-1} R = I. \tag{127}$$

A set of these properties is called a "group." The set of rotations (126) is a group.

In discussing this group, we begin by considering a point p in space with coordinates (x, y, z). If we rotate the space, point p will now have new coordinates (x', y', z'). If we knew the rotation R we could represent the new coordinates (x', y', z') in terms of (x, y, z) as follows:

$$x' = \alpha_{11} x + \alpha_{12} y + \alpha_{13} z$$

$$y' = \alpha_{21} x + \alpha_{22} y + \alpha_{23} z \tag{128}$$

$$z' = \alpha_{31} x + \alpha_{32} y + \alpha_{33} z,$$

or in matrix notation

$$X' = AX \tag{129}$$

where X' and X are column vectors, and $A = (\alpha_{ij})$ is a matrix. Since the distance from a point p to the origin must remain constant during rotation, we have

$$x^2 + y^2 + z^2 = x'^2 + y'^2 + z'^2, \tag{130}$$

which implies that

$$\alpha_{11}^2 + \alpha_{21}^2 + \alpha_{31}^3 = 1^2 + 0^2 + 0^2 = 1 \tag{131}$$

and similar formulas for the other α's.

Let X^T (a row vector) be the transpose of X; similarly for X'^T and X'. The transpose of A, obtained by interchanging rows and columns, is A^T. The length of X' is the square root of the scalar product $X'^T X' = (AX)^T AX = X^T (A^T A) X$. Since length is preserved in rotation, $X'^T X' = X^T X$ for all vectors X, we must have

$$A^T A = I \tag{132}$$

where I is the identity matrix. A matrix A which satisfies this relation is called an *orthogonal* matrix. We have shown that every rotation matrix is an orthogonal matrix.

The converse is not true. There are orthogonal matrices which are not rotation matrices; for example, the reflection matrix

$$
\begin{pmatrix}
-1 & 0 & 0 \\
0 & -1 & 0 \\
0 & 0 & -1
\end{pmatrix}
$$

although orthogonal, is not a rotation.

Consider the problem of finding all rotation matrices. The relation (132) states that

$$
\begin{pmatrix}
\alpha_{11} & \alpha_{21} & \alpha_{31} \\
\alpha_{12} & \alpha_{22} & \alpha_{32} \\
\alpha_{13} & \alpha_{23} & \alpha_{33}
\end{pmatrix}
\begin{pmatrix}
\alpha_{11} & \alpha_{12} & \alpha_{13} \\
\alpha_{21} & \alpha_{22} & \alpha_{23} \\
\alpha_{31} & \alpha_{32} & \alpha_{33}
\end{pmatrix}
=
\begin{pmatrix}
1 & 0 & 0 \\
0 & 1 & 0 \\
0 & 0 & 1
\end{pmatrix}
\quad (133)
$$

which leads to quadratic relations such as

$$
\alpha_{11}^2 + \alpha_{21}^2 + \alpha_{31}^2 = 1 ,
$$

$$
\alpha_{11}\alpha_{12} + \alpha_{21}\alpha_{22} + \alpha_{31}\alpha_{32} = 0 ,
$$

etc. There are nine relations for the nine unknown α_{ij}, but these relations are not all independent. We shall show how these quadratic relations can be replaced by linear relations.

Consider a one-parameter subgroup of rotations $A(t)$, such that for each value of t we have

$$
A^T(t)A(t) = I \tag{134}
$$

$$
A(0) = I , \tag{135}
$$

and

$$
A(t_1)A(t_2) = A(t_1 + t_2) . \tag{136}
$$

We shall use calculus arguments to investigate the structure of the group $A(t)$. Since $A(0) = I$, we write for $\Delta t \approx 0$,

$$
A(\Delta t) \approx I + \Delta t S , \tag{137}
$$

where S is independent of t. Then

$$[A(\Delta t)]^T = I + \Delta t S^T . \tag{138}$$

Using (132) for the rotation matrix $A(\Delta t)$, we find

$$A A^T = (I + \Delta t S)(I + \Delta t S^T) = I + \Delta t(S + S^T) + O[(\Delta t)^2] = I + O[(\Delta t)^2] \tag{139}$$

provided that

$$S + S^T = 0 .$$

This requires that S be a skew-symmetric matrix:

$$S^T = -S . \tag{140}$$

Now let us find $dA(t)/dt$. We have

$$\frac{dA(t)}{dt} = \lim_{\Delta t \to 0} \frac{A(t + \Delta t) - A(t)}{\Delta t} \tag{141}$$

and because of (136), $A(t + \Delta t) = A(t)A(\Delta t)$. Thus

$$\frac{dA(t)}{dt} = \lim \left[\frac{A(\Delta t)A(t) - A(t)}{\Delta t} \right] = \lim \frac{A(\Delta t) - I}{\Delta t} A(t) = SA(t) \tag{142}$$

from (137). The solution of the differential equation

$$\frac{dA}{dt} = SA(t) \tag{143}$$

is

$$A(t) = e^{tS} . \tag{144}$$

This discussion has shown that the rotation matrix (A) is the exponential of a skew-symmetric matrix (S). What is the most general skew-symmetric matrix? If we put

$$\begin{pmatrix} a & b & c \\ d & e & f \\ g & h & i \end{pmatrix} = - \begin{pmatrix} a & d & g \\ b & e & h \\ c & f & i \end{pmatrix}$$

we find that $a = -a$, $b = -d$, etc.; consequently, the most general form of a skew-symmetric matrix is

$$\begin{pmatrix} o & b & c \\ -b & o & f \\ -c & -f & o \end{pmatrix}. \tag{145}$$

We write this result as

$$S = bS_3 - cS_2 + fS_1,$$ (146)

where

$$S_1 = \begin{pmatrix} 0 & 0 & 0 \\ 0 & 0 & 1 \\ 0 & -1 & 0 \end{pmatrix}, \quad S_2 = \begin{pmatrix} 0 & 0 & -1 \\ 0 & 0 & 0 \\ 1 & 0 & 0 \end{pmatrix}, \quad S_3 = \begin{pmatrix} 0 & 1 & 0 \\ -1 & 0 & 0 \\ 0 & 0 & 0 \end{pmatrix}.$$ (147)

It is clear that any skew-symmetric matrix can be written as a linear combination of S_1, S_2, S_3. Since a rotation is the exponential of a skew-symmetric matrix, we conclude that every rotation matrix can be written as

$$e^{a_1 S_1 + a_2 S_2 + a_3 S_3}.$$ (148)

Thus the rotation group is a three-parameter group; this means that a rotation can be defined by specifying only three quantities, for example, the Euler angles of the rotation. Note that in terms of matrix multiplication, we have $S_1 S_2 - S_2 S_1 = -S_3$, etc., thus

$$[S_1, S_2] = -S_3,$$

$$[S_2, S_3] = -S_1,$$ (149)

$$[S_3, S_1] = -S_2.$$

These relations show that the skew-symmetric matrices are elements of a Lie algebra. The study of this Lie algebra is simpler than the study of the corresponding Lie group of rotations, because as we have seen, the S's are defined by only three parameters instead of by the nine apparently involved in the rotation matrix. The S's are called the *infinitesimal generators* of the rotation group.

CHAPTER 7

Probability

7.1 INTRODUCTION

In Chapters 2 and 4 we introduced various special topics of probability theory in order to illustrate applications of other subjects. Now we consider certain aspects of probability more systematically. As a general reference for this chapter, we use *Modern Probability Theory and Its Applications* by E. Parzen.

To be able to discuss the subject of probability adequately, we must become familiar with the technical language. We therefore start with a list of definitions.

Sample - the outcome of an experiment or observation. For example, consider the experiment of drawing two cards from a deck of cards. Suppose a King and a Queen are drawn. Our sample is indicated as (K, Q).

Sample Description Spaces - the total set of possible outcomes of the experiments. For example, if we consider a short deck of cards containing only A, 2, 3, . . ., 10 of one suit and we are taking two card samples, the sample description space would be the following array of possible pairs:

$$(A, 2)(2, 3) \ldots (9, 10)$$

$$(A, 3)(3, 4) \ldots$$

$$\cdot$$
$$\cdot$$
$$\cdot$$

$$(A, 10)$$

If the experiment consists of drawing one card, replacing it, and then drawing a second card, the sample description space would also contain samples such as (A, A). Thus the sample description space is the set

$$S = \left\{ (i, j): \ 1 \le i \le 10, \ 1 \le j \le 10, \ i \ne j \right\},$$

if we sample without replacement, and is the set

$$S = \left\{ (i, j): \ 1 \le i \le 10, \ 1 \le j \le 10 \right\},$$

if we sample with replacement.

Event - a set of specified samples (a set of descriptions of samples). An event E is a subset of S - the sample description space. S is itself an event; it is called the "certain event."

THE ALGEBRA OF SETS

Suppose E and F are sets of samples. The *union* of E and F, written $E \cup F$, is the set of samples which belong to either E or F. The *intersection* of E and F, written $E \cap F$, is the set of samples which belong to both E and F.

The operations of forming the union and intersection satisfy certain rules; they are

1. Commutative

$$E \cup F = F \cup E; \quad E \cap F = F \cap E$$

2. Associative

$$(E \cup F) \cup G = E \cup (F \cup G); \quad (E \cap F) \cap G = E \cap (F \cap G).$$

3. Distributive

$$E \cap (F \cup G) = (E \cap F) \cup (E \cap G)$$

and

$$E \cup (F \cap G) = (E \cup F) \cap (E \cup G)$$

Two sets are equal if the sets are identical. We may interpret the sets E, F, $E \cap F$, and $E \cup F$ as point sets by using Figure 7.1.

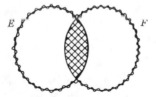

Figure 7.1

Here E and F correspond to the individual ovals, $E \cap F$ to the overlapped, crosshatched region, and $E \cup F$ to the region inside the curve singled out by the wavy overlay. Similarly, in Figure 7.2,

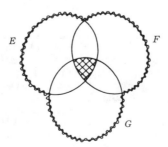

Figure 7.2

the region corresponding to $(E \cap F) \cap G$ is crosshatched, and the curve bounding $(E \cup F) \cup G$ is marked by a wavy line. It is possible to have a situation is which E and F do not intersect, in which case $E \cap F$ is empty. This is indicated by $E \cap F = \emptyset$. The set $\emptyset$ is called the "impossible set" (no samples). If $E \cap F = \emptyset$, then E and F are *mutually exclusive* or *disjoint*. See Figure 7.3.

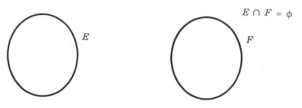

$$E \cap F = \phi$$

$$E \qquad F$$

Figure 7.3

The *complement* of a set E is the set of samples which do not belong to E. This is indicated by $\overline{E}$. Note that

$$E \cap \overline{E} = \emptyset$$

$$E \cup \overline{E} = S \text{ (the whole space } S\text{)}.$$

Suppose we have events E and F and we want to write an expression for combinations of the events E and F. The event in which both E and F occur is $E \cap F$. The event in which neither occur is $\overline{E} \cap \overline{F}$. The event in which at least one of the events occur is $E \cup F$. The event in which at most one of the events occur is $\overline{E \cap F}$. Finally, the event in which exactly one of the events E or F occur is $(E \cap \overline{F}) \cup (\overline{E} \cap F)$.

Notice that $E \cup E = E$ and $E \cap E = E$. Also, the complement of $\overline{E}$ is $\overline{\overline{E}} = E$.

DeMorgan's law states that the complement of a union is the intersection of complements and that the complement of an intersection is the union of complements, that is, $\overline{E \cup F} = \overline{E} \cap \overline{F}$, and $\overline{E \cap F} = \overline{E} \cup \overline{F}$.

The notation $F \subset E$ means that F is a *subevent* of E; that is, all samples in F are in E. If $F \subset E$ and $E \subset F$, then $F = E$.

FINITE SAMPLE SPACE

We define $N[E]$ as the number of samples in E. Note that if E and F are disjoint, then $N[E + F] = N[E] + N[F]$. As an example of a problem dealing with the number of elements of a set, let us look at a survey taken by a magazine. The survey found that in the group of 1000 subscribers there were 312 males, 470 married persons, 525 college graduates, 42 male college graduates, 147 married college graduates, 86 married males, and 25 married male college graduates. At first these figures seem to be quite reasonable, but we can prove that these figures are inconsistent. Indicating male by the letter M, married by R, college graduate by K, and the entire field of subscribers by S, we can set up the following equation:

$$S = S \cap S \cap S = (M \cup \overline{M}) \cap (R \cup \overline{R}) \cap (K \cup \overline{K}) .$$

Expanding S into its eight disjoint classes, we get

$$S = (M \cap R \cap K) \cup (M \cap R \cap \overline{K}) \cup (M \cap \overline{R} \cap K) \cup (M \cap \overline{R} \cap \overline{K}) \cdots .$$

Since the individual sets are disjoint, we find

$$N[S] = N(M \cap R \cap K) + N(M \cap R \cap \overline{K}) + \cdots .$$

Dropping $\cap$ for brevity, we have

$$N[MRK] = 25, \quad N[\overline{M}\overline{R}K] = 361$$

$$N[MR\overline{K}] = 61, \quad N[\overline{M}R\overline{K}] = 262$$

$$N[\overline{M}RK] = 122, \quad N[M\overline{R}\overline{K}] = 209$$

$$N[M\overline{R}K] = 17, \quad N[\overline{M}\overline{R}\overline{K}] = 0$$

The total of these is 1057, not 1000, which proves that the figures of the survey are inconsistent.

7.2 ELEMENTARY PROPERTIES OF PROBABILITY

Probability is a real-valued function of events; its value for each event E is written $P[E]$, and it has the following properties:

1. $P[E] \geq 0$

2. $P[S] = 1$

3. $P[E \cup F] = P[E] + P[F]$ if $E \cap F = \emptyset$.

From these properties one finds the following to be true:

$$P[\emptyset] = 0, \quad P[S \cup \emptyset] = P[S] + P[\emptyset] = P[S] .$$

Since

$$F = S \cap F = (E \cup \overline{E}) \cap F = (E \cap F) \cup (\overline{E} \cap F)$$

and these sets are disjoint, we have

$$P[F] = P[E \cap F] + P[\overline{E} \cap F] ;$$

therefore

$$P[F \cap \overline{E}] = P[F] - P[E \cap F] .$$

Note that if $F = S$, we get

$$P[\overline{E}] = 1 - P[E] .$$

The probability is always between 0 and 1.

For the union of three disjoint sets E_1, E_2, E_3 (that is, if $E_1 \cap E_2 = \emptyset$, $E_2 \cap E_3 = \emptyset$, $E_3 \cap E_1 = \emptyset$), we have

$$P[E_1 \cup E_2 \cup E_3] = P[E_1] + P[E_2] + P[E_3] .$$

To see why this is true, look at

$$(E_1 \cup E_2) \cap E_3 = (E_1 \cap E_3) \cup (E_2 \cap E_3) \quad = \emptyset \cup \emptyset = \emptyset$$

$$P[E_1 \cup E_2] = P[E_1] + P[E_2]$$

$$P[E_1 \cup E_2 \cup E_3] = P[E_1 \cup E_2] + P[E_3]$$

$$= P[E_1] + P[E_2] + P[E_3] .$$

METHODS OF ASSIGNING PROBABILITY

One may obtain a probability by assigning values to the samples in a sample description space, provided that the sum of these values for the entire space is equal to one. For example, say we toss three coins. Our sample description space is made up of eight possible samples:

$$(HHH), (HHT), (HTH), (HTT), (THH), (THT), (TTH), (TTT). \qquad (*)$$

Assuming that each of these elementary samples is equally likely, we would assign the value 1/8 to each sample. The probability of the event that exactly two heads occur is the probability of the event which contains the three samples (HHT), (HTH), and (THH). Because this event is the union of three disjoint events, namely, the event that HHT occurs, the event that HTH occurs, and the event that THH occurs, we see that

$$P[E] = P[(HHT)] + P[(HTH)] + P[(THH)]$$

$$= 1(1/8) + 1(1/8) + 1(1/8) = 3/8 .$$

We shall now introduce the concept of a "*random variable.*" A random variable is a real-valued function on S (the sample description space). As an example of the use of the random variable in a problem, let us say we are gambling upon the toss of three coins. For every head that turns up, we win one dollar, for every tail, we lose one dollar. Let $f(s)$ represent the value of the random variable for the sample s. Then, for the samples (HHT) and (TTT), we would win or lose thus;

$$f(HHT) = \$1 + \$1 - \$1 = \$1$$

$$f(TTT) = -\$1 - \$1 - \$1 = -\$3 .$$

EXPECTED VALUES

We define the *expectation* of the random variable $f(s)$ as

$$E(f) = \sum f(s)P[(s)] .$$

The possible values of $f(s)$ for the samples (HHH), (HHT), (HTH), etc., are

$$+3, +1, -1, \text{ and } -3 \ .$$

The expectation of f [ordering the samples as indicated in $(*)$] is

$$E(f) = 3(1/8) + 1(1/8) + 1(1/8) - 1(1/8) + 1(1/8) - 1(1/8) - 1(1/8) - 3(1/8)$$

or

$$E(f) = 3(1/8) + 1(3/8) - 1(3/8) - 3(1/8) = 0 \ .$$

Note that the last line is an illustration of the formula

$$E(f) = \sum_s kP[\{s:f(s) = k\}] \ .$$

If we have two random variable functions f and g, the expectation of their sum is

$$E(f + g) = \sum [f(s) + g(s)]P[s] = \sum f(s)P[(s)] + \sum g(s)P[(s)]$$

$$= E(f) + E(g) \ .$$

Suppose that f is as already described, that is, \$1 is lost for a tail and \$1 is won for a head, whereas g is the random variable which describes winning \$1 if the first coin of the sample is a head and losing \$1 if the first coin of the sample is a tail. We can see that $f + g$ is the sum of two random variables and is itself a random variable. The possible values of $f + g$ for the samples $(*)$ are 4, 2, 0, -2, -4, which have the corresponding probabilities $1/8, 2/8, 2/8, 2/8, 1/8$. The expectation of $f + g$ is then

$$E(f + g) = 4(1/8) + 2(2/8) + 0(2/8) - 2(2/8) - 4(1/8) = 0 \ ,$$

that is,

$$E(f) + E(g) = 0 \ .$$

Another problem which we may solve by using the concept of random variables is one in which we want to find, say, the expected number of heads that will occur if we toss n coins. The sample space of this problem consists of n H's and T's or 2^n elements. We do not assume that the events are equally likely. If we assign the value p to be the probability of a head falling, then the probability of a tail is $1 - p$. Define random variables f_k, $k = 1, 2, \ldots, n$ as follows:

$$f_k = \begin{cases} 1 \text{ if the } k\text{th coin is } H \ , \\ 0 \text{ if the } k\text{th coin is } T \ . \end{cases}$$

Note that $E(f_k) = p$ because $1p + 0(1 - p) = p$. If $t = f_1 + f_2 + \cdots + f_n$, the random variable t counts 1 for each H and 0 for each T; thus t is the number of heads which occurs in the sample of n tosses. We want to find the expected value of the random variable t. We have

$$E(t) = \sum E(f_k) = \sum p = np \ .$$

Since

$$P[\{s:t(s) = k\}] = \binom{n}{k} p^k (1 - p)^{n-k} , \qquad \binom{n}{k} = \frac{n!}{(n-k)!k!} ,$$

we see that

$$E(t) = \sum k \binom{n}{k} p^k (1 - p)^{n-k} = np \ .$$

INDICATOR FUNCTIONS

An *indicator function* of an event G is the random variable defined as follows:

$$I_G(s) = \begin{cases} 1 \text{ if } s \text{ is in the event } G , \\ 0 \text{ if } s \text{ is not in the event } G . \end{cases}$$

We will now show some of the properties of the indicator function. Suppose we have two indicator functions $I_G(s)$ and $I_F(s)$. These functions interact in the following manner:

$$I_{G \cap F} = I_G \times I_F$$

$$I_{G \cup F} = I_G + I_F \quad \text{if} \quad G \cap F = \emptyset$$

$$E(I_F) = 1 \cdot P[\{s:I_F(s) = 1\}] + 0 \cdot P[\{s:I_F(s) = 0\}] = P[F] \ .$$

We will now solve a sample problem using the methods thus far described. If a man's secretary has n letters and n envelopes and she randomly places a letter in each envelope, let us find the probability that no letter will be placed in the correct envelope. We will call the sample description space S and the event of the kth letter being placed in the correct envelope A_k. We use f_k to denote the indicator function for A_k; that is, $f_k = 1$ if the kth letter is in the correct envelope, otherwise, $f_k = 0$. The indicator function for the event that no letter goes into the correct envelope is

$$I = (1 - f_1)(1 - f_2)(1 - f_3) \cdots (1 - f_n) \ .$$

We have

$$E\left((1 - f_1)(1 - f_2) \cdots (1 - f_n)\right)$$

$$= E\left(1 - \sum_{k=1}^{n} f_k + \sum_{k_1=1}^{n} \sum_{k_2 > k_1}^{n} f_{k_1} f_{k_2} - \sum_{k_1 < k_2 < k_3}^{n} f_{k_1} f_{k_2} f_{k_3} + \cdots + (-1)^n f_1 f_2 \cdots f_n\right)$$

$$= 1 - \sum E(f_k) + \sum \sum E(f_{k_1} f_{k_2}) + \cdots + (-1)^n E(f_1 \cdots f_n)$$

Since

$$E\left(f_k\right) = E\left(I_{A_k}\right) = P\left[A_k\right] \ ,$$

and

$$E\left(f_{k_1} f_{k_2}\right) = E\left(I_{A_{k_1} \cap A_{k_2}}\right) = P\left[A_{k_1} \cap A_{k_2}\right] \ ,$$

we see that

$$P[\text{no correct letters}] = 1 - \sum_{k=1}^{n} P\left[A_k\right] + \sum P\left[A_{k_1} \cap A_{k_2}\right]$$

$$- \sum P\left[A_{k_1} \cap A_{k_2} \cap A_{k_3}\right] + \cdots$$

$$+ (-1)^n P\left[A_{k_1} \cap A_{k_2} \cap \cdots\right] \ .$$

Here, $P\left[A_k\right]$ indicates the probability that the kth letter went into the correct envelope; since all letters and envelopes are the same, we have

$$P\left[A_k\right] = \frac{1}{n} = \frac{(n-1)!}{n!} \ .$$

The term $P\left[A_{k_1} \cap A_{k_2}\right]$ indicates the probability that two letters went into the correct envelopes, and this is

$$P\left[A_{k_1} \cap A_{k_2}\right] = \frac{1}{n} \frac{1}{n-1} = \frac{(n-2)!}{n!} \ .$$

The probability that all the letters went into the correct envelopes is

$$P\left[A_{k_1} \cap A_{k_2} \cap \cdots \cap A_{k_n}\right] = \frac{1}{n} \frac{1}{n-1} \frac{1}{n-2} \cdots \frac{1}{1} = \frac{1}{n!} \ .$$

The probability that no letters went into the correct envelopes is

$$P[\text{no events}] = 1 - n \frac{(n-1)!}{n!} + \binom{n}{2} \frac{(n-2)!}{n!} - \binom{n}{3} \frac{(n-3)!}{n!}$$

$$+ \cdots + \frac{(-1)^n}{n!}$$

$$= 1 - \frac{1}{1!} + \frac{1}{2!} - \frac{1}{3!} + \frac{1}{4!} - \cdots + \frac{(-1)^n}{n!}$$

$$\sim e^{-1} \approx 0.37 \ .$$

DEPENDENT EVENTS

We now wish to consider the case in which we have two related events, A and B. What is the probability of B assuming that A has already occurred? When a condition (such as the previous occurrence of A) is placed upon the problem of finding the probability of an event B, we have a problem of conditional probability. This is indicated by $P[B \mid A]$ and is defined as $P[A \cap B]/P[A]$, if $P[A] \neq 0$. If $P[A] = 0$, then $P[B \mid A]$ is undefined. If we consider a set of equally likely elementary events, then

$$P[A] = \frac{N[A]}{N[S]} = \frac{\text{number of successes}}{\text{number of possible samples}}$$

and

$$P[A \cap B] = \frac{N[A \cap B]}{N[S]} \; ;$$

therefore

$$P[B \mid A] = \frac{N[A \cap B]}{N[A]} \quad .$$

Again using the case of tossing three coins as an example, we will find the probability of obtaining exactly two H's out of the sample space (*) assuming that the first coin tossed comes up a head. The possible samples are those in which the first coin comes up H, that is, (HHH), (HHT), (HTH), and (HTT). Since exactly two heads occur in only two of these samples, we get

$$P[A \mid B] = \frac{N[A \cap B]}{N[A]} = \frac{P[A \cap B]}{P[A]} = \frac{2}{4} \quad .$$

For another example, consider a family which has two children. What is the probability that both children will be boys if (1) we know that the older child is a boy, or (2) we know that at least one child is a boy? Will the probability be the same for both case (1) and (2)? Suppose that B is the event that the family has two boys, A_1 that the older child is a boy, and A_2 that one child is a boy. Note that $P[A_1] = 1/2$, $P[A_2] = 3/4$, and $P[B] = 1/4$. Since $P[A_1 \cap B] = P[A_2 \cap B] = 1/4$, we see that

$$P\left[B \mid A_1\right] = \frac{1/4}{1/2} = 1/2,$$

and

$$P\left[B \mid A_2\right] = \frac{1/4}{3/4} = 1/3 \quad .$$

INDEPENDENT EVENTS

With this background, we now introduce the concept of "independence." If $P[B \mid A] = P[B]$, then A and B are *independent*. From the previous discussion, the probability $P[B \mid A] = P[A \cap B]/P[A]$; but if A and B are independent, then

$$P[A \cap B] = P[A]P[B] \text{ and } N[A \cap B] = N[A]N[B]/N[S].$$

Similarly, we extend the development directly to any number of events which are independent of each other. Let us consider three events A, B, and C, and suppose that C is independent of A and B. We have

$$P[C \mid A \cap B] = \frac{P[A \cap B \cap C]}{P[A \cap B]} = P[C]$$

and

$$P[A \cap B \cap C] = P[A \cap B]P[C] .$$

If A and B are also independent, we have

$$P[A \cap B \cap C] = P[A]P[B]P[C]$$

and

$$P[A \cap B] = P[A]P[B], \; P[B \cap C] = P[B]P[C], \; P[A \cap C] = P[A]P[C] .$$

It is also possible for any pair of events to be independent, but for the group of all three events not to be independent. As an example, say we have a deck of four cards $(\mathcal{A}, 2, 3, 4)$, from which we draw samples of one card. Let $A = \{\mathcal{A} \text{ or } 2\}$, $B = \{\mathcal{A} \text{ or } 3\}$, and $C = \{\mathcal{A} \text{ or } 4\}$. Now we will show that any two of these events are independent, but not all three. We have $P[A] = 1/2$, $P[B] = 1/2$, $P[C] = 1/2$, and $P[A \cap B] = P[\mathcal{A}] = 1/4 = (1/2)(1/2) = P[A]P[B]$; thus A and B are independent. Similarly, we can show that any other pair of events are independent. But if we consider all three events simultaneously, we find that the three events are not independent of one another because

$$P[A \cap C \cap B] = P[\mathcal{A}] = 1/4 \neq P[A]P[B]P[C] = 1/8 .$$

Thus, in this example we find that the knowledge of the occurrence of one event does not change the probability of the other events, but knowing the outcome of two of the three events does change the probability of the third, i.e., $P[C \mid A] = P[C] = 1/2$ but $P[C \mid A \cap B] = 1$.

If we are given the dependent events $A_1, A_2, \cdots, A_n$, we find that the conditional probability of the event A_n, given that A_{n-1}, $A_{n-2}, \cdots, A_1$ have occurred, is

$$\left[P \, A_n \mid A_{n-1} \cap A_{n-2} \cap \cdots \cap A_1 \right] = \frac{P[A_n \cap A_{n-1} \cap \cdots \cap A_1]}{P[A_{n-1} \cap A_{n-2} \cap \cdots \cap A_1]} .$$

The probability of the joint occurrence of these events is

$$P\left[A_n \cap A_{n-1} \cap \cdots \cap A_1\right]$$

$$= P\left[A_{n-1} \cap A_{n-2} \cap \cdots \cap A \,\middle|\, P\left[A_n \,\middle|\, A_{n-1} \cap \cdots \cap A_1\right]\right.$$

$$= P\left[A_1\right]P\left[A_2 \,\middle|\, A_1\right]P\left[A_3 \,\middle|\, A_1 \cap A_2\right] \cdots P\left[A_n \,\middle|\, A_1 \cap \cdots \cap A_{n-1}\right] .$$

PRODUCT SPACES

When speaking of sample description spaces, we must use the appropriate mathematical framework. There is a conceptual distinction between, say, tossing each of n coins once, or tossing one coin n times.

If we have a sample space consisting of one coin with the possibility of two outcomes (H) or (T), which we write as $S = \{(H), (T)\}$, we may form a sample description space for two trials with one coin;

$$S_1 \times S_2 = \{(H) \times (H), (H) \times (T), (T) \times (H), (T) \times (T)\} ,$$

or a sample description space for three trials with one coin,

$$S_1 \times S_2 \times S_3 = \{(H) \times (H) \times (H), (H) \times (H) \times (T), \cdots\}$$

and so on for any number of trials. In this way we see that we have formed a group of product sample description spaces. In this particular example, the spaces S_1, S_2, etc., are over the same experiment, but the spaces could also be formed on different, unrelated experiments, e.g., S_1 could be the experiment of tossing a coin, S_2 could be throwing a pair of dice, and so on.

In a like manner, we may form a product event. If C_1 is an event in S_1, and C_2 is an event in S_2, the product event $C_1 \times C_2$ is

$$C_1 \times C_2 = \{(s) \times (t): \ s \text{ in } C_1; \ t \text{ in } C_2\} .$$

Suppose we have one die, which we toss twice. The individual sample spaces are

$$S_1 = \{(1), (2), \ldots, (6)\} ;$$

$$S_2 = \{(1), (2), \ldots, (6)\} .$$

The product space is

$$S_1 \times S_2 = \{(i) \times (j): \ 1 \leq i \leq 6, \ 1 \leq j \leq 6\} .$$

If we let C_1 indicate the event that an even number appears in the first experiment and C_2 indicate the event that a number less than 3 appears in the second experiment, we have the product event

$$C_1 \times C_2 = \{(2) \times (1), (2) \times (2), (4) \times (1), (4) \times (2), (6) \times (1), (6) \times (2)\}.$$

Not every event in a product space is a product event. In the above case the product event is a subset within the product space.

We now define the probability of the product event, knowing the probability of the separate events. Using the above example, let us find the probability of the product event $C_1 \times C_2$. We have

$$P_1[C_1] = 1/2, \qquad P_2[C_2] = 1/3$$

and then

$$P[C_1 \times C_2] = P_1[C_1]P_2[C_2] = 1/6 \ .$$

We will next discuss the problem of finding the probability of an event such as

$$S_1 \times S_2 \times \cdots \times S_{k-1} \times C_k \times S_{k+1} \times \cdots \times S_n \ ,$$

which is defined as the kth sample space of a product sample description space. Say we have three sample spaces: S_1 = tossing a coin with samples $(H)(T)$; S_2 = throwing a die with samples $(1)(2) \cdots (6)$; and S_3 = drawing a card from a deck. Let C represent the event in S_3 which corresponds to drawing a picture card $(J)(Q)(K)$ from the deck.

The product space is

$$S_1 \times S_2 \times S_3 = \{i, j, k: \ i = (H), (T), j = (1), (2), \ldots, (6), k = \text{card}\} \ .$$

The desired event is

$$S_1 \times S_2 \times C = \{i, j, k: i = (H), (T), j = (1), (2), \ldots, (6), k = (J), (Q), (K)\}.$$

An event depending upon the kth trial is independent of an event depending upon the jth trial. Let A be an event in S_2 which indicates tossing an even number on a die. The event that depends upon the kth trial and the jth trial is written as

$$(S_1 \times S_2 \times C) \cap (S_1 \times A \times S_3) = S_1 \times A \times C \ .$$

The probability of each of these events is

$$P[S_1 \times S_2 \times C] = P_1[S_1]P_2[S_2]P_3[C] = P_3[C]$$

$$P[S_1 \times A \times S_3] = P_1[S_1]P_2[A]P_3[S_3] = P_2[A]$$

$$P[S_1 \times A \times C] = P_1[S_1]P_2[A]P_3[C] = P_2[A]P_3[C] \ .$$

As an example of our methods, consider the game "Odd Man Out."
In this game, n people each toss a coin. If the face of one man's coin
is different from all the others, he is out of the game. Let S_k with
$k = 1, 2, \ldots n$, be the sample space of the kth player's coin, that is, let
$S_k = \{(H), (T)\}$. The space of the game is $S_1 \times S_2 \times \cdots \times S_n$.
We are interested in finding the probability that there will be $(n - 1)$ H's
and one T, or $(n - 1)$ T's and one H. The probability is $n(1/2)^n$ for one
H and the rest T, $n(1/2)^n$ for one T and the rest H. These events are
mutually exclusive; thus the probability of one man going out of the game
is $2n(1/2)^n = n/2^{n-1}$; for a specific example, if $n = 5$, the probability
of one man going out is $5/16$.

Another problem is to find the probability of one man staying in the
game for a specified number of trials. We now set up a new space to find
the probability of one man going out on, say, the seventh toss and not
before. Let T_1 be the one space with the events f = man does not go out,
and s = man does go out. Then the space is

$$T_1 \times T_2 \times \cdots \times T_6 \times T_7.$$

For five players as above, for which case

$$P[f] = 11/16, \quad P[s] = 5/16,$$

we have

$$P[(f) \times (f) \times \cdots \times (s)] = (11/16)^6 (5/16) .$$

7.3 INFINITE SAMPLE SPACES

So far, we have discussed only finite sample spaces. The extension
to infinite sample spaces raises questions on what we mean by counting
and integration. We shall discuss the elementary aspects of these topics,
but merely sketch the rest.

To enable us to find the probability of events in an infinite space we
must be able to assign a probability to the individual samples (s) of the
infinite sample space $\{s_1, s_2, s_3, \ldots\}$. If p_j is the probability of (s_j),
we require that

$$p_j \geq 0; \quad j = 1, 2, 3, \ldots, \quad \text{and} \quad \sum_1^\infty p_j = 1 .$$

If these conditions are met we can manipulate probabilities on infinite
spaces much as we did on finite spaces, e.g., $P[\{s_1\} \cup \{s_2\}]$
$= P[\{s_1\}] + P[\{s_2\}] = p_1 + p_2$.

These methods, through adequate for simple infinite sample spaces,
are not sufficient to deal with the problems which arise when the sample
space is not countable, i.e., when the samples cannot be paired with the
sequence of positive integers.

Let us look at an example of more complicated infinite sample
spaces.

Say we toss a coin an infinite number of times. We get an infinite ordered sequence of heads and tails. If we toss a second coin an infinite number of times, we get a different ordered sequence. By the methods that we have discussed so far, we cannot assign a probability to each sample, because there are too many to be counted.

COUNTING

As a preliminary we briefly discuss the idea of "counting" in connection with infinite sets. George Cantor developed the concept of one-to-one correspondence as a tool for determining whether two infinite sets have the same number of elements. The explanation of this concept is as follows: If, given two sets S and T, for every sample (s) in S there is a sample (t) in T and if for every sample (t) in T there is a sample (s) in S, then S and T are said to be in one-to-one correspondence. This concept can be used for both finite and infinite sets.

As an example of two sets which are in one-to-one correspondence, let us look at (1) the set of all odd numbers, 1, 3, 5, 7, . . . , $2n - 1$, . . . , and (2) the set of all square numbers, 1, 4, 9, 16, 25, . . . , n^2, ... We set up a correspondence between the samples in each set. The correspondence can be arbitrary (the numbers need not be equal to each other), but there must be one element in the first set which corresponds to each element in the second set and one element in the second set corresponding to each element in the first set. In our example the correspondence would be as follows: 1 ⟷ 1, 3 ⟷ 4, 5 ⟷ 9, 7 ⟷ 16, ... Thus the two sets (1) and (2) are in one-to-one correspondence.

If we are accustomed to thinking in terms of finite sets, the concept of one-to-one correspondence is difficult to understand intuitively. In the above example, it would appear intuitively that the set (1) would have more elements than the set (2), but this is not true because the series are infinite.

As another example of how this concept may give results which are intuitively paradoxical, we show that the number of points on a line one unit long is equal to the number of points on a line two units long. This is proved by setting up a one-to-one correspondence between each point on the one unit line and each point on the second line. In Figure 7.4 for every point such as P on the one-unit line, there is a point P' on the two-unit line:

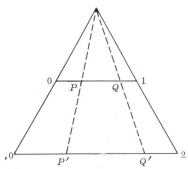

Figure 7.4

Thus we have set up a one-to-one correspondence between every point on the one-unit line and every point on the two-unit line; therefore the number of points on a line one unit long is equal to the number of points on a line two units long.

Consider another problem. Can there be as many rational numbers as there are integers? To show that there are as many fractions as integers, we must be able to put the two sets in one-to-one correspondence. The integers are already ordered, but we need to be able to put the set of all fractions into an order which not only includes all fractions but which can be put into one-to-one correspondence with the integers. We can do this, if we arrange the fractions in a coordinate array, as shown in Figure 7.5.

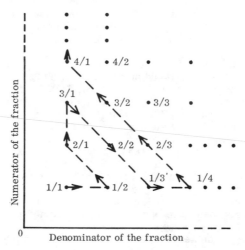

Figure 7.5

Following the pattern indicated, the fractions can be ordered to include all possibilities, and after eliminating duplications, to enable us to set up a one-to-one correspondence between the fractions and the integers. Note that an infinite set can be placed in one-to-one correspondence with a subset of that set if and only if the subset itself is infinite.

A set that can be put in one-to-one correspondence with the sequence of positive integers is called "denumberable" or countable.

Theorem: A countable set of countable sets is countable.

Suppose we have the countable sets

$$S_1 = (a_1, a_2, a_3, a_4, \cdots)$$

$$S_2 = (b_1, b_2, b_3, \cdots)$$

$$S_3 = (c_1, c_2, c_3, \cdots)$$

$$S_4 = (d_1, d_2, \cdots)$$

$$\cdot$$
$$\cdot$$
$$\cdot$$

The infinite set which is the union of these infinite sets will still be a countable set because the elements may be ordered (or counted) in the the manner shown in Figure 7.6:

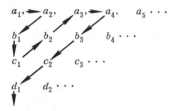

Figure 7.6

In addition to the infinite sets which are countable, there are very many which are uncountable, e.g., the set of all points on the interval zero to one or the set of all infinite decimals.

The fact that the set of all infinite decimals is not countable is not intuitively clear. To prove that the set is not countable, let us assume that we can find a one-to-one correspondence between the set of infinite decimals and the integers. We will then show that this cannot be done, thus proving by contradiction that the set of infinite decimals cannot be counted.

Assume that the set of decimals below is the ordered set of all infinite decimals:

$$0.a_1 a_2 a_3 a_4 a_5 \cdots$$
$$0.b_1 b_2 b_3 b_4 b_5 \cdots$$
$$0.c_1 c_2 c_3 c_4 c_5 \cdots$$
$$0.d_1 d_2 d_3 d_4 d_5 \cdots$$

However, there is at least one additional decimal formed from the diagonal digits a_1, b_2, c_3, $\ldots$ by adding 1 to each digit except when a digit is 9, in which case we substract 1. Thus we get a decimal such as $0.(a_1 + 1)(b_2 + 1)(c_3 + 1) \cdots$. This decimal is not in the above set because it differs from the kth decimal at least in the kth digit. Thus, we find that we cannot arrange the infinite decimals in an order so that they may all be included and counted.

The set of all sequences of heads and tails obtained when a coin is tossed an infinite number of times has the same property of being uncountable. Suppose some sequences are

$$(HH\,TT\,H\,TT \cdots)$$
$$(H\,TT\,T\,H\,TH \cdots)$$
$$(H\,TH\,H\,H\,TT \cdots)$$
$$(TH\,T\,T\,H\,H\,T \cdots)$$

.
.
.

To each sequence of tosses we can assign a decimal value, e.g., $(HTHHT\cdots)$ can be assigned the number

$$\frac{1}{10} + \frac{2}{10^2} + \frac{1}{10^3} + \frac{1}{10^4} + \frac{2}{10^5} + \cdots$$

where each H corresponds to 1 in the numerator and each T to 2. This shows we have the same problem with the sequences of tosses as with the infinite decimals.

There is a simpler and more useful method of assigning numbers to the sequences. If we let the H's be represented by 1's and the T's represented by 0's, we may set up binary decimals to represent the values of the sequences, e.g., $(HTHHT\cdots)$ could be represented by

$$\frac{1}{2} + \frac{0}{2^2} + \frac{1}{2^3} + \frac{1}{2^4} + \frac{0}{2^5} + \cdots$$

which would be written in binary notation as $0.10110\ldots$. Using the binary notation, we can investigate limits, such as $\lim_{n\to\infty}(H_n/n)$ where H_n is the number of heads in the first n tosses of a particular sequence.

MEASURE

When we attempt to assign probability to an uncountably infinite set (such as the number of points on a line), we find that it is undesirable to set up a probability for each individual point because the summation of an uncountable sequence does not make sense.

To set up probabilities for uncountable sample spaces, we impose the same conditions as for the previous problems, namely,

S = sample space, $E \subset S$ (E = subset of S)

(1) $P[E] \geq 0$

(2) $P[S] = 1$

(3) $P[E \cup F] = P[E] + P[F]$,   if $E \cap F = \emptyset$

and assign numbers to the sets of points. But first, it becomes necessary to generalize the additivity property (3) to the property (3′) of complete additivity:

(3′) probability of an infinite union

$$P[E_1 \cup E_2 \cup E_3 \cup \cdots] = \sum_1^\infty P[E_j]$$

if every two sets are disjoint, $E_j \cap E_k = \emptyset$.

For example, if we consider a space S to be the points on a line one unit long, and E_1, E_2, etc., to be segments of that line, i.e., subsets of the space S, we can assign probabilities to the events $E_1, E_2, \ldots$, even though the events are subsets of an uncountable set. We shall use the indicator function for these events, that is, the function

$$\varphi_E(s) = \begin{cases} 1 \text{ if } s \text{ is in } E, \\ 0 \text{ if } s \text{ is not in } E \end{cases}.$$

We define the probability of an event E as $\int_0^1 \varphi_E(s)ds$. Thus if E is the interval from a to b we have

$$P[E] = \int_0^1 \varphi_E(s)ds = \int_a^b 1\,ds = b - a,$$

and for two disjoint events $E(a$ to $b)$ and $F(c$ to $d)$, we have

$$P[E \cup F] = \int_0^1 \varphi_{E \cup F}(s)ds = \int_a^b ds + \int_c^d ds = (b - a) + (d - c);$$

our definition of probability satisfies the first three properties above.

If around every point in a set T, there is some interval completely contained in T, then T is called an *open set*. An interval with no end points is an open set. Also, the union of a countable number of sets which have no end points is an open set. Every open set can be written as a countable union of disjoint open intervals I_k, so, if we have an open set E, the probability of that set may be defined as $P[E] = \Sigma P[I_k]$.

If we define the *measure* of a set as

$$P[E] = \int_0^1 \varphi_E(s)ds = \text{measure of } E,$$

the measure of an interval is equal to its length. We find that not every set can be measured, that is, there exist nonmeasurable sets.

To illustrate the concept of measurable sets, let us find the measure of the set of rational points. The set of rational points is not an open set and not an interval. We can cover the set of rational points by an open set. If E is an open set and R represents the rational numbers

$$R \subset E, \quad \text{then } P[R] \le P[E].$$

We shall close in on $P[R]$ by limiting E. Knowing that the rational numbers are countable, we let ϵ be the length of an interval around the first rational point r_1, $\epsilon/2$ the length of the interval around the second rational point r_2, $\epsilon/2^2$ the length of the interval around the third rational point r_3, and so on. We find

$$0 \le P[R] \le P[E] \le \epsilon + \epsilon/2 + \epsilon/2^2 + \epsilon/2^3 + \cdots = \epsilon \sum 1/2^n = 2\epsilon.$$

Thus the measure of the rational points is smaller than any ϵ we pick, and we can say that the measure of the set of rational points is zero.

INTEGRATION

The measure of a set E is often indicated by $\mu(E)$. It has the following properties:

(1) $\mu(E) \geq 0$

(2) $\mu(\overset{\infty}{\underset{1}{\cup}} E_n) = \sum \mu(E_n)$ if $E_i \cap E_j = \emptyset$.

Consider the three sets

(1) $\{x: x$ rational, $0 < x < 1\}$. This set has no largest or smallest rational number, but this set is countable.

(2) $\{x: x$ irrational, $0 < x < 1\}$.

(3) $\{x: x$ real, $0 < x < 1\}$. This set is the union of set (1) and set (2). We have shown that this set is not countable.

We know that the union of a countable number of countable sets is countable. Since we know that set (3) is uncountable and that set (3) is the union of the countable set (1) with the set (2), we conclude set (2) must be uncountable. We have shown that the measure of set (1) is zero, and the measure of set (3) (the unit line) is one; the measure of set (2), the difference of (3) and (1), is thus also one.

We need to have a method of integrating functions of the type

$$r(x) = \begin{cases} 0 \text{ if } x \text{ is rational} \\ 1 \text{ if } x \text{ is irrational} \end{cases}$$

so that we may find the expectation of such events. An analytic representation of a function of the type of $r(x)$ is

$$\lim_{m \to \infty} \lim_{n \to \infty} (1 - \cos 2\pi m! x)^n = \begin{cases} 0 \text{ if } x \text{ rational} \\ 1 \text{ if } x \text{ irrational} . \end{cases}$$

Before mentioning the role of measure in integrating functions such as $r(x)$, let us discuss the basic definition of an integral.

Suppose we have a function defined over the interval $(0, 1)$ by $y = f(x)$. We divide that interval into smaller intervals to find an approximation of the integral $\int_0^1 f(x) dx$; thus corresponding to Figure 7.7,

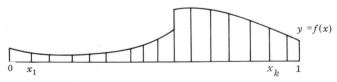

Figure 7.7

we see that

$$\sum_{0}^{n-1} f(x_k)(x_{k+1} - x_k)$$

is an approximation of the integral. Letting the length of the intervals go to zero, we get

$$\lim \sum_{0}^{n-1} f(x_k)(x_{k+1} - x_k) = \int_0^1 f(x)\,dx,$$

which is the elementary definition of an integral. But it is well known that this definition does not apply to functions of the type of $r(x)$; every interval contains rational and irrational numbers, and the value of the limit depends on how we choose the subdivisions, e.g., we get zero if the x_k are rational and unity if the x_k are irrational.

Let us try another approach. We can approximate $f(x)$ by step functions $S_1(x)$ and $s_1(x)$, as shown in Figure 7.8.

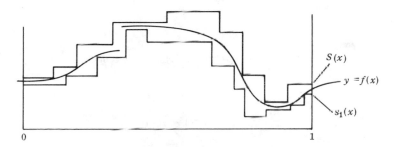

Figure 7.8

We assume that we know the integrals of the approximating step functions:

$$s_1(x) \le f(x) \le S_1(x)$$

$$\int s_1(x)\,dx \le \int f(x)\,dx \le \int S_1(x)\,dx .$$

We continue to take better approximations $S_1, S_2, \ldots,$ and $s_1, s_2, \ldots,$ always keeping $f(x)$ between the two step functions:

$$\int s_n(x)\,dx \le \int f(x)\,dx \le \int S_n(x)\,dx .$$

If we can find sequences of step functions such that the limit of the difference of the integrals of the two step functions equals zero, i.e.,

$$\lim_{n \to \infty} [\int_0^1 S_n(x)\,dx - \int_0^1 s_n(x)\,dx] = 0 ,$$

then $f(x)$ is integrable, and the integral of $f(x)$ is defined as

$$\int f(x)\,dx = \lim \int s_n(x)\,dx = \lim \int S_n(x)\,dx \ .$$

To be sure that this definition is valid, we must show that $\int f(x)\,dx$ is not dependent upon our choice of step functions. Let $T_n(x)$ and $t_n(x)$ indicate step functions different from $S_n(x)$ and $s_n(x)$, where

$$t_n(x) \le f(x) \le T_n(x)$$

and

$$\lim \int t_n(x)\,dx = \lim \int T_n(x)\,dx = \tau \ .$$

we must have $\tau = \int f(x)dx$ because

$$t_n(x) \le f(x) \le S_n(x)$$

and

$$\int t_n(x)\,dx \le \int S_n(x)\,dx \ ;$$

therefore

$$\tau = \lim \int t_n(x)\,dx \le \lim \int S_n(x)\,dx = \int f(x)\,dx \ .$$

Also

$$s_n(x) \le f(x) \le T_n(x)$$

and

$$\int f(x)\,dx = \lim \int s_n(x)\,dx \le \lim \int T_n(x)\,dx = \tau \ .$$

Since $\tau \le f(x)dx$ and $\tau \ge \int f(x)dx$, it follows that $\tau = \int f(x)dx$.

Essentially, this definition of an integral is the same as the Riemann integral and will not hold for the $r(x)$ type of function.

We can also define a double integral by a similar method, using rectangular solids as we did step functions. Say we have the volume $z = f(x, y)$ which we approximate by $s_1(x, y)$ and $S_1(x, y)$, whose integrals $\int \int s_1\,dx\,dy$ and $\int \int S_1\,dx\,dy$ are known See Figure 7.9. Using a sequence of s's and S's, we approximate $f(x, y)$:

$$s_1 \le s_2 \le s_3 \le \cdots \le f(x, y) \le S_1 \le S_2 \le S_3 \cdots .$$

If $\lim [\int \int S_n\,dx\,dy - \int \int s_n\,dx\,dy] \to 0$, then (x, y) is integrable and

$$\int\int f(x, y)dx \, dy = \lim \int \int S_n dx \, dy = \lim \int \int s_n \, dx \, dy .$$

This definition of a double integral is again independent of our choice of the approximating functions s_n and S_n.

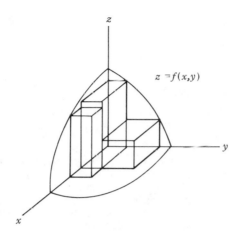

$z = f(x,y)$

Figure 7.9

We now wish to show that a double integral as defined above is the same as a repeated integral, that is,

$$\int dx[\int f(x, y) \, dy] = \int \int f(x, y)dx \, dy .$$

Consider the function $f(x, y)$, where x is fixed; then

$$s_n(x, y) \le f(x, y) \le S_n(x, y) .$$

We know that $\int \int S_n(x, y) \, dx \, dy = \int dx[\int S_n(x, y)dy]$. Put $T_n = \int S_n(x, y)dy$ and $t_n(x) = \int s_n(x, y)dy$. If $\lim_{n\to\infty}[T_n - t_n] = 0$, then we can integrate f with respect to y, because

$$\int f(x, y)dy = \lim t_n(x) = \lim T_n(x) .$$

If $\lim \int [T_n(x) - t_n(x)]dx = 0$, then $\int f(x, y)dy$ can be integrated with respect to x to give

$$\int dx[\int f(x, y)dy] = \lim \int t_n(x)dx = \lim \int dx \int s_n(x, y)dy = \int \int f(x, y)dx \, dy .$$

However, the assumption that the double integral exists implies only that

$$\lim [\int \int S_n dx \, dy - \int \int s_n \, dxdy] = \lim [\int T_n dx - \int t_n \, dx] = 0 .$$

If also $\lim(T_n - t_n) = 0$, we can be sure that the repeated integral will equal the double integral. But it is not always true that the limit of an integrand is zero when the limit of the integral is zero. For an illustration of this, consider the sequence of function $\Psi_n(x)$ defined to equal unity or zero on different lengths of the interval $(0, 1)$:

Ψ_1
$$\frac{1 \quad \mid \quad 0}{0 \qquad\qquad 1} \, , \qquad \Psi_2 \quad \frac{0 \quad \mid \quad 1}{0 \qquad\qquad 1} \, , \qquad \Psi_3 \quad \frac{1 \mid 0 \mid 0 \mid 0}{0 \qquad\qquad\quad 1} \, ,$$

Ψ_4
$$\frac{0 \mid 1 \mid 0 \mid 0}{0 \qquad\qquad\quad 1} \, , \qquad \Psi_5 \quad \frac{0 \mid 0 \mid 1 \mid 0}{0 \qquad\qquad\quad 1} \, , \qquad \Psi_6 \quad \frac{0 \mid 0 \mid 0 \mid 1}{0 \qquad\qquad\quad 1} \, , \text{ etc.}$$

For this sequence of functions, the limit never converges to zero, but the limit of the integral does converge to zero; that is, $\lim \Psi_n(x)$ does not exist but $\lim \int \Psi_n(x)\,dx = 0$.

This difficulty is remedied by the Lebesgue integral. The Lebesgue definition of integration also applies to $r(x)$ of page 224. The essential feature of the new integration procedure is that we ignore values of the integrand on sets whose measure is zero. We say that a set that can be covered by an open set of arbitrarily small length (as illustrated on page 223) is a set of measure zero. Thus the function $r(x)$, which is zero for the rationals (a set of measure zero) and unity for the irrationals on the unit line (a set of measure one) has the integral one over the unit line. Similarly, for a function $z = f(x, y)$, it can be shown that except for a set of measure zero in x, $f(x, y)$ is integrable with respect to x. Every countable set is a set of measure zero because every countable set can be covered by an open set which can be made arbitrarily small.

An interesting example of a set of measure zero is Cantor's "middle third" set. Divide the interval from zero to one into thirds. Take away the middle third. Next, divide the segments which are left into thirds and take away the middle thirds. If we continue this indefinitely, we have left what is called Cantor's middle third set. A number whose ternary expansion contains only 0 and 2, such as 0.200220002022..., will be a point in the set. Let us find the measure of the set of all numbers whose ternary expansion has no digit 1. The part of the unit interval which we took away has the total length

$$\frac{1}{3} + \frac{2}{9} + \frac{4}{27} + \frac{8}{81} + \cdots = \frac{1/3}{1 - (2/3)} = 1$$

Since we have taken away intervals whose length totals one, we have left a set of measure zero even though the set has an infinite uncountable number of elements.

Let us see how we can measure more general sets E. Cover the set E by the open set O whose measure is $\mu(O)$. We have

$$O - E = \{x:\ x \text{ in } O, \text{ but } x \text{ not in } E\} \, .$$

We want the set $O - E$ to be small. If we can find a sequence of open sets

O_n, such that an open set u_n covers $O_n - E$ and such that $\mu(u_n) \to 0$, then we say E is measurable, and the measure of E is $\mu(E) = \lim \mu(O_n)$. This idea is due to Lebesque.

If the sets E and F are measurable, then their union, $E \cup F$, their intersection $E \cap F$, and their complements $\overline{E}$ and $\overline{F}$, are all measurable. The function $f(x)$ is measurable if $\{x: f(x) < a\}$ is measurable for all values of a. We can assign probability to measurable sets. We cannot assign probability to nonmeasurable sets.

Every bounded measurable function can be integrated. Every bounded measurable function $f(x)$ can be approximated by a simple function $s_n(x)$, and the integral of $f(x)$ is the limit of the integral of the simple function:

$$\int f(x)dx = \lim \int s_n(x)dx .$$

The types of convergence which we may expect from the function $s_n(x)$ are as follows:

(1) point-wise convergence, $s_n(x) \to f(x)$ for every value of x;

(2) mean-square convergence, $s_n(x) \to f(x)$ if $\int |f(x) - s_n(x)|^2 dx \to 0$;

(3) convergence almost everywhere, $s_n(x) \to f(x)$ almost everywhere (except for a set of measure zero);

(4) convergence in measure, $s_n(x) \to f(x)$ in measure $\mu\{\text{set } |f(x) - s_n(x)| > \epsilon\} \to 0$.

Convergence almost everywhere implies convergence in measure, but convergence in measure does not imply convergence almost everywhere.

If we have an abstract space corresponding to a sample description space, then: (1) only a certain class of events can be assigned probabilities, or equivalently, (2) only a certain class of events can be measured. A class of sets A_n for which A_n is measurable, $\overline{A}_n$ is measurable, $\cap A_n$ is measurable, and $\cup A_n$ is measurable, is called a sigma field (σ field) of sets.

The three components, S (a space), $\mathcal{A}$ (a σ field of sets), and P (the measure of the sets in $\mathcal{A}$) is called a normalized measure space if it has the following properties:

$$P[S] = 1$$

$$P[\phi] = 0$$

$$1 \geq P[A] \geq 0$$

$$P[\cup A_n] = P[A_1] + P[A_2] + \cdots, \qquad A_m \cap A_n = \phi.$$

our probability corresponds to the usual Lebesgue measure.

We correlate some terms used in the discussion of measure spaces to those of probability spaces:

sample probability space	normalized measure space
elementary event	element in space
event	measure space in $\mathcal{A}$
certain event	S
impossible event	$\emptyset$
probability of $A = P[A]$	measure of $\ = \mu(A)$
almost surely	almost everywhere
random variable F	bounded measurable function F
expected value of F	Lebesgue integral of F
limit in probability	limit in measure
limit almost surely	limit almost everywhere

If F is a random variable, the expectation $E(F)$ is $\int F(s)\,dP(s)$. To illustrate this, let F_n be a simple function such that

$$F_n = \sum_1^n f_k I_{E_k}, \qquad \int F_n\,dP = \sum f_k \int I_{E_k}\,dP = \sum f_k P[E_k].$$

We have lim (in measure) $F_n = F$, and

$$\int F\,dP = \lim \int F_n\,dP = \lim \sum f_k P[E_k].$$

Put

$$P[\{s:\ F(s) < x\}] = D(x)$$

where $D(x)$ = distribution function of the random variable F. Thus, the expectation of F is

$$E(F) = \int_{-\infty}^{\infty} x\,dD(x) = \int_{-\infty}^{\infty} xp(x)\,dx.$$

where $p(x)$, the probability density, can be introduced whenever $D(x)$ is differentiable.

CHAPTER 8

Perturbation Theory

Since most differential equations cannot be solved in closed form, it is important to have techniques available that can be used to approximate the solutions. One of the most useful of such techniques is perturbation from the solution of a simpler equation which can be solved exactly. In the first section we shall consider the cases where the standard perturbation theory is applicable. In later sections we shall consider cases where the standard theory breaks down and so-called singular perturbation methods must be used.

8.1 REGULAR PERTURBATION THEORY

For the standard perturbation theory, it is just as easy and slightly more general to consider arbitrary operators instead of differential operators.

Consider an equation depending on a small positive parameter ϵ,

$$(L - \epsilon M)u = f \tag{1}$$

where L and M are given operators, the function f is known, and the function u must be determined. We shall assume that $Lu = f$ can be solved explicitly by using the known inverse operator L^{-1} of L. Applying L^{-1} to (1), we get

$$u - \epsilon L^{-1}Mu = L^{-1}f$$

which we rewrite, with obvious simplifications, as

$$u = \epsilon Ku + g . \tag{2}$$

Equation (2) will be solved by a standard iteration method. We assume some function u_0 is an approximation to the solution, put it in the right-hand side, and get (hopefully) a better approximation

$$u_1 = g + \epsilon Ku_0 .$$

If u_1 is not a close enough approximation, we try again and get

$$u_2 = g + \epsilon Ku_1 .$$

This procedure can be continued indefinitely by using the recurrence relation

$$u_{n+1} = g + \epsilon K u_n; \quad n = 0, 1, 2, \cdots; u_0 = g . \tag{3}$$

This procedure is meaningful if, first, we can show that the sequence u_n of approximations converges to a limit, and secondly, that this limit is the solution of (2).

The meaning of convergence depends on what *norm* we use, that is, on what measure of closeness we use. For example, we may say $\varphi(x)$ is close to zero if

$$\max_{0 \le x \le 1} |\varphi(x)| \tag{4}$$

or if

$$\int_0^1 |\varphi(x)|^2 dx \tag{5}$$

is close to zero. Both (4) and (5) are examples of norms. More precisely, the *norm* of a function $\varphi(x)$ in the interval $0 \le x \le a$ is a number, denoted by the symbol $\|\varphi\|$, such that

(a) $\|\varphi\| \ge 0$, and $\|\varphi\| = 0$ implies $\varphi = 0$;
(b) $\|\alpha\varphi\| = |\alpha| \cdot \|\varphi\|$, if α is a constant;
(c) $\|\varphi + \psi\| \le \|\varphi\| + \|\psi\|$, for arbitrary functions φ and ψ.

It is clear that (4) and (5) satisfy (a), (b), and (c).

Let us return to (3) and consider the difference of successive approximations. We have

$$u_{n+1} - u_n = \epsilon K u_n - \epsilon K u_{n-1} , \tag{6}$$

where the operator K need not be linear. Taking the norm of (6), we get

$$\| u_{n+1} - u_n \| = \epsilon \| K u_n - K u_{n-1} \| . \tag{7}$$

Now, we assume that

$$\| K u_n - K u_{n-1} \| \le C \| u_n - u_{n-1} \| , \tag{8}$$

where C is a constant independent of n. Put $\rho_{n+1} = \| u_{n+1} - u_n \|$; then because of (7) and the assumption about K we find

$$\rho_{n+1} < \epsilon C \rho_n$$

for all values of n. This implies that

$$\rho_{n+1} < \epsilon C \rho_n < \epsilon C (\epsilon C \rho_{n-1}) < \cdots < (\epsilon C)^n \rho_1 . \tag{9}$$

If $\epsilon C < 1$, this inequality shows that ρ_{n+1}, the norm of the difference between successive approximations, goes to zero as n goes to infinity.

This result does not imply convergence. To discuss convergence, consider the infinite series

$$u_0 + (u_1 - u_0) + (u_2 - u_1) + \cdots . \tag{10}$$

Since the $(n + 1)$th partial sum of the series is u_n, the infinite series and the sequence u_n will either both converge or both diverge. Convergence here means convergence in the sense of the norm.

By taking the norm of the remainder of the series after the nth term and using (9), we find that

$$\| (u_{n+1} - u_n) + (u_{n+2} - u_{n+1}) + \cdots \| \le \rho_{n+1} + \rho_{n+2} + \cdots \tag{11}$$

$$< \sum_{k=0}^{\infty} (\epsilon C)^{n+k} \rho_1 = \frac{(\epsilon C)^n}{1 - \epsilon C} \rho_1 .$$

Thus, if $\epsilon C < 1$, the norm of this remainder converges to zero as n goes to infinity; consequently, the series and the sequence converge to a limit function v, say. Now,

$$\lim K u_n - K v + \lim \lfloor K u_n - K v \rfloor ,$$

and because of the assumption (8),

$$\lim \| K u_n - K v \| \le \lim C \| u_n - v \| = 0 ;$$

that is, the right-hand side equals zero because u_n converges to v; consequently $\lim K u_n = K v$. Using this fact when we go to the limit in (3), we find

$$v = g + \epsilon K v .$$

Therefore v is the solution of (2). Hereafter, we will denote this solution by the symbol u as before.

From (11) we can estimate the closeness of the approximation u_n to u. Since $u - u_n = (u_{n+1} - u_n) + (u_{n+2} - u_{n+1}) + \cdots$, we find that

$$\| u - u_n \| < \frac{(\epsilon C)^n}{1 - \epsilon C} \| u_1 - u_0 \| . \tag{12}$$

This formula enables us to estimate the error of an approximation even without knowing the exact solution.

We shall illustrate these ideas by applying them to the solution of

$$y' = f(x, y) \tag{13}$$

with the boundary condition $y(0) = y_0$. The function $f(x, y)$ need not be linear. We regard (13) as the form $Ly = \epsilon My$ with

$$Ly = y', \qquad \epsilon My = f(x, y) .$$

Since the inverse of L is integration, we get

$$y = y_0 + \int_0^x f[\xi, y(\xi)] \, d\xi . \tag{14}$$

The sequence of approximations is generated by using

$$y_{n+1}(x) = y_0 + \int_0^x f[\xi, y_n(\xi)] \, d\xi ,$$

and the difference of successive approximations is obtained from

$$y_{n+1}(x) - y_n(x) = \int_0^x \{f[\xi, y_n(\xi)] - f[\xi, y_{n-1}(\xi)]\} \, d\xi . \tag{15}$$

Let us define the norm as

$$\| \varphi \| = \max_{0 \le x \le a} |\varphi(x)| .$$

We have from (15)

$$\| y_{n+1}(x) - y_n(x) \| \le \int_0^x |f[\xi, y_n(\xi)] - f[\xi, y_{n-1}(\xi)]| \, d\xi .$$

Let us assume

$$\left| \frac{\partial f}{\partial y} \right| < C$$

for $0 \le x \le a$, and for all y. Since

$$|f[\xi, y_n(\xi)] - f[\xi, y_{n-1}(\xi)]| = \left| \int_{y_{n-1}}^{y_n} \frac{\partial f(\xi, \eta)}{\partial y} \, d\eta \right| < C |y_n(\xi) - y_{n-1}(\xi)| ,$$

we have

$$|y_{n+1}(x) - y_n(x)| < C \int_0^x |y_n(\xi) - y_{n-1}(\xi)| \, d\xi . \tag{16}$$

Put

$$\rho_{n+1}(x) = |y_{n+1}(x) - y_n(x)| ;$$

then (16) implies that

$$\rho_{n+1}(x) < C \int_0^x \rho_n(\xi) \, d\xi .$$

We must now estimate $\rho_n(x)$ and show that $\sum_n \rho_n(x)$ converges.

Let us consider the more general inequality

$$\rho_{n+1}(x) < \int_0^x \rho_n(\xi) m(\xi) \, d\xi \,, \tag{17}$$

where $m(\xi)$ is a positive function, and let us try to obtain an estimate for ρ_{n+1} which does not depend on ρ_n. To do this, we shall assume, for a moment, that the inequality (17) is actually the equality

$$\rho_{n+1}(x) = \int_0^x \rho_n(\xi) m(\xi) \, d\xi \,, \tag{18}$$

and we shall also assume that $\rho_0(x)$ is equal to a constant B_0, say. Writing

$$\int_0^x m(\xi) \, d\xi = M(x) \,,$$

we find from the integral equation,

$$\rho_1(x) = B_0 M(x) \,,$$

$$\rho_2(x) = B_0 \int_0^x M(\xi) m(\xi) \, d\xi = B_0 \int_0^x M(\xi) \, dM(\xi) - B_0 (M^2/2!) \,,$$

and by induction

$$\rho_n(x) = B_0 \frac{[M(x)]^n}{n!} \,.$$

Similarly for the inequality (17), we assume, by analogy with the solution of (18), that

$$\rho_n(x) = B_n(x) \frac{[M(x)]^n}{n!} \,,$$

and (17) becomes

$$B_{n+1}(x) \frac{[M(x)]^{n+1}}{(n+1)!} < \int_0^x B_n(\xi) \frac{M(\xi)^n}{n!} m(\xi) \, d\xi < \|B_n\| \frac{M(x)^{n+1}}{(n+1)!} \,.$$

Thus $B_{n+1}(x) \le \|B_n\|$, and it follows that $|B_{n+1}(x)| < B_0$ for all values of n and $0 \le x \le a$; consequently, (17) implies

$$\rho_n(x) < B_0 \frac{[M(x)]^n}{n!} \,, \quad \text{if} \quad \rho_0(x) < B_0 \quad \text{for } 0 \le x \le a.$$

Returning to the case $m(\xi) = C$ in (16), we find that the sequence $y_n(x)$ in (15) satisfies the inequality

$$\left| y_{n+1}(x) - y_n(x) \right| < B_0 \frac{C^n x^n}{n!} \,, \tag{19}$$

where

$$| y_0(x) | < B_0 .$$

The estimate (19) enables us to show that (13) has a solution for $0 \le x \le a$ because the infinite series

$$y_0 + (y_1 - y_0) + (y_2 - y_1) + \cdots$$

is dominated by the convergent series

$$B_0 + B_0 Cx + B_0 \frac{C^2 x^2}{2!} + \cdots = B_0 e^{Cx} ,$$

and therefore converges for $0 \le x \le a$. This shows that the iteration process converges to a limit, and by the same methods as before we can show that the limit is the solution of (13).

8.2 THE FREDHOLM EXPANSION

Let us consider again the problem of finding a solution of the equation $u + \epsilon Ku = g$, if K is now a linear operator. For example, consider the equation

$$u(x) + \epsilon [(1 - x) \int_0^x \xi u(\xi) d\xi + x \int_x^1 (1 - \xi) u(\xi) d\xi] = g , \qquad (20)$$

where g is a given function of x. We put

$$u_0 = g$$

$$u_1 = g - \epsilon Ku_0 = g - \epsilon Kg ,$$

and by induction we get the Neumann series

$$u = [I - \epsilon K + (\epsilon K)^2 - (\epsilon K)^3 + \cdots]g ,$$

where

$$\epsilon Kg = \epsilon [(1 - x) \int_0^x \xi g(\xi) d\xi + x \int_x^1 (1 - \xi) g(\xi) d\xi] ,$$

and the higher terms require successive integrations.

In this form it is not obvious for which values of ϵ the expansion converges. Let us change the integral equation (20) to a differential equation by differentiating twice. Thus we have

$$u''(x) + \epsilon u(x) = g''(x)$$

with the conditions $u(0) = g(0)$ and $u(1) = g(1)$. To show what difficulties may arise, suppose we let $g(x) = x$, so that the equation becomes

$$u'' + \epsilon u = 0, \quad u(0) = 0, \quad u(1) = 1 .$$

The solution of this is $u = \sin x \epsilon^{\frac{1}{2}} / \sin \epsilon^{\frac{1}{2}}$, but trouble will occur when $\sin \epsilon^{\frac{1}{2}} = 0$, that is, for $\epsilon = \pi^2$. It can be shown that the power series breaks down and no solution exists for $\epsilon = \pi^2$ but that the series will converge for $|\epsilon| < \pi^2$.

We now discuss a method which will give an expansion of a solution of $u = g + \epsilon K u$ for all ϵ for which the solution exists, and which will make obvious the values of ϵ for which solutions do not exist. Consider the formal expansion of the inverse of

$$(I - \epsilon K)u = g ; \tag{21}$$

that is,

$$u = \frac{1}{I - \epsilon K} g = [1 + (\epsilon K) + (\epsilon K)^2 + \cdots]g ,$$

where I is the identity operator. Were ϵK a number, then we would require $|\epsilon K| < 1$ for the power series to converge; were $\epsilon K = 1$, difficulties would arise with the denominator of $1/(I - \epsilon K)$. A similar difficulty occurs in the solution of a system of m linear equations, for which case I and K are n by n matrices, and u and g are column vectors: the solution breaks down when the determinant of $I - \epsilon K$ is zero. Let us continue to the general case by analogy as if we knew what $\det(I - \epsilon K)$ meant.

We write

$$u = \frac{1}{1 - \epsilon K} g = \left[1 + \epsilon \frac{D(K, \epsilon)}{D(\epsilon)}\right] g , \quad D(\epsilon) = \det(I - \epsilon K) , \tag{22}$$

where $D(\epsilon)$ is a scalar (that is, not an operator), and where we isolate the operator $D(K, \epsilon)$ from $(1 - \epsilon K)[1 + \epsilon D(K, \epsilon)/D(\epsilon)] = I$ in the form

$$D(K, \epsilon) = \frac{D(\epsilon)K}{1 - \epsilon K} . \tag{23}$$

Were K a finite-dimensional matrix, then (23) would suggest that $D(K, \epsilon)$ is regular for all values of ϵ, singularities of the numerator cancelling those of the denominator. From this we surmise that both $D(K, \epsilon)$ and $D(\epsilon)$ have expansions in powers of ϵ which converge for all ϵ. To obtain these expansions, we must first define the determinant of a general operator. We use the matrix case as a guide.

If M is a matrix with eigenvalues λ_j (such that $Mu = \lambda_j u$), then the determinant of M is the product of the eigenvalues:

$$\det M = \prod_j \lambda_j = \lambda_1 \lambda_2 \cdots \lambda_n . \tag{24}$$

Since

$$\log \det M = \sum_{j=1}^{n} \log \lambda_j = \text{trace} \log M, \tag{25}$$

where "trace" stands for the sum of the diagonal elements of a matrix (here of the matrix $\log M$ whose eigenvalues are $\log \lambda_j$), we obtain

$$\det M = e^{\text{trace} \log M} \tag{26}$$

This formula can be used to define "determinant" even if M is not a matrix, provided we interpret "trace" appropriately. Thus, we write $\det (I - \epsilon K) = \exp[\text{trace} \log (I - \epsilon K)]$ and

$$\log D(\epsilon) = \log [\det(I - \epsilon K)] = \text{trace} \log (I - \epsilon K), \tag{27}$$

where if

$$Ku = \int_{b}^{a} K(x, \xi)u(\xi)d\xi, \tag{28}$$

then

$$\text{trace } K = \int_{a}^{b} K(\xi, \xi)d\xi. \tag{29}$$

We could expand $D(\epsilon)$ in a power series by means of the formal representation

$$\log (I - \epsilon K) = -\sum_{1}^{\infty} \frac{\epsilon^n K^n}{n},$$

from which

$$\text{trace} \log (I - \epsilon K) = -\sum \frac{\epsilon^n}{n} \text{trace } K^n. \tag{30}$$

However, there are more convenient routes based on starting with sums with unknown coefficients.

Thus we assume the series expansion

$$D(\epsilon) = \sum_{0}^{\infty} \frac{(-1)^n C_n}{n!} \epsilon^n, \tag{31}$$

where $C_0 = 1$ and the C_n are constants, and similarly

$$D(K, \epsilon) = \sum_{0}^{\infty} \frac{(-1)^n B_n}{n!} \epsilon^n \tag{32}$$

where the B_n are operators. We now substitute these series into

$$(I - \epsilon K)D(K, \epsilon) = KD(\epsilon),$$

as obtained from (23), and get

$$(I - \epsilon K)\sum \frac{(-1)^n B_n}{n!}\epsilon^n = K\sum \frac{(-1)^n C_n}{n!}\epsilon^n ,$$

from which

$$B_0 = KC_0 = K ; \quad \frac{(-1)^n B_n}{n!} + (-1)^n \frac{KB_{n-1}}{(n-1)!} = \frac{(-1)^n KC_n}{n!} , n > 0 .$$

We now have

$$B_0 = KC_0 = K ; \quad B_n + nKB_{n-1} = KC_n , n > 0 , \tag{33}$$

but this is a recurrence relation involving both B's and C's. We must now find another relation between the two sets.

To obtain an additional relation, we differentiate (27) with respect to ϵ,

$$\frac{D'(\epsilon)}{D(\epsilon)} = - \text{trace} \frac{K}{I - \epsilon K} , \tag{34}$$

and then substitute from (23) into the right-hand side:

$$\frac{D'(\epsilon)}{D(\epsilon)} = - \text{trace} \frac{D(K, \epsilon)}{D(\epsilon)} .$$

Cancelling $D(\epsilon)$, we have

$$D'(\epsilon) = - \text{trace} D(K, \epsilon) , \tag{35}$$

and substituting the series (31) and (32) we obtain

$$\sum_1^\infty \frac{(-1)^n C_n}{(n-1)!}\epsilon^{n-1} = -\sum_0^\infty (-1)^n \frac{\text{trace } B_n}{n!}\epsilon^n .$$

Thus for $n > 0$,

$$C_n = \text{trace } B_{n-1} \tag{36}$$

which together with (33) enables us to find all successive values of B_n and C_n from $C_0 = 1$.

Working out the first few terms, and introducing $k_n = \text{trace } K^n$, we get

$$C_0 = 1 , \qquad B_0 = K ; \tag{37}$$

$$C_1 = k_1 , \qquad B_1 = k_1 K - K^2 ;$$

$$C_2 = k_1^2 - k_2 , \qquad B_2 = (k_1^2 - k_2)K - 2k_1 K^2 + 2K^3 ;$$

and so on. To obtain explicit results for given K's, we would substitute (37) into (22) in terms of (31) and (32), that is, into

$$u = g + \epsilon \, \frac{B_0 - B_1\epsilon + B_2\epsilon^2/2! + \cdots}{1 - C_1\epsilon + C_2\epsilon^2/2 + \cdots} \, g \, . \tag{38}$$

This formula is known as Fredholm's expansion for the solution of (21). This should converge except at the eigenvalues where the denominator vanishes, that is, except where "resonances" occur.

In particular, if K is an integral operator as in (28), then

$$Kg = \int K(x, \, \xi)g(\xi) \, d\xi \, ,$$

where $K(x, \, y)$ is called the kernel of the integral operator K; as before, in (29) for trace K, we have

$$k_1 = \int K(\xi, \, \xi) \, d\xi \, .$$

Similarly

$$K^2 g = K(Kg) = \int K(y, \, x) \, dx \int K(x, \, \xi)g(\xi) \, d\xi$$

$$= \int [\, \int K(y, \, x)K(x, \, \xi) \, dx]g(\xi) \, d\xi, \tag{39}$$

so that $\mathrm{trace}\, K^2$ equals

$$k_2 = \int \int K(\xi, \, x)K(x, \, \xi) \, dx \, d\xi, \tag{40}$$

etc. Writing $B_n(x, \, y)$ for the kernel of the integral operator B_n, we convert (33) to

$$B_n(x, \, y) = -n \int K(x, \, z)B_{n-1}(z, \, y) \, dz + K(x, \, y)C_n \, , \tag{41}$$

and (36) to

$$C_n = \int B_{n-1}(y, \, y) \, dy \, . \tag{42}$$

The present results for B_n can be written in terms of integral operators whose kernels are determinants. Thus the corresponding forms of (37) are

$$C_0 = 1 ,$$

$$B_0(t_0, t_1) = K(t_0, t_1) ;$$

$$C_1 = \int K(t, t) \, dt ,$$

$$B_1(t_0, t_2) = \int \begin{vmatrix} K(t_0, t_2) & K(t_0, t_1) \\ K(t_1, t_2) & K(t_1, t_1) \end{vmatrix} dt_1$$

$$= \int [K(t_0, t_2)K(t_1, t_1) - K(t_0, t_1)K(t_1, t_2)] \, dt_1 \qquad (43)$$

$$= K(t_0, t_2) \int K(t_1, t_1) \, dt_1 - \int K(t_0, t_1)K(t_1, t_2) \, dt_1 ;$$

$$C_2 = \int B_1(t, t) \, dt = \left[\int K(t, t) \, dt \right]^2 - \int \int K(t, t_1)K(t_1, t) \, dt \, dt_1 ,$$

$$B_2(t_0, t_3) = \int \int \begin{vmatrix} K(t_0, t_3) & K(t_0, t_1) & K(t_0, t_2) \\ K(t_1, t_3) & K(t_1, t_1) & K(t_1, t_2) \\ K(t_2, t_3) & K(t_2, t_1) & K(t_2, t_2) \end{vmatrix} dt_1 \, dt_2 , \text{ etc.}$$

Perturbation of Eigenvalues. We now discuss a perturbation technique for obtaining the eigenvalues of a linear operator $L = L_0 + \epsilon L_1$ when we know the eigenvalues and eigenfunctions of L_0. We assume L_0 to be self-adjoint.

To solve

$$(L_0 + \epsilon L_1)u = \lambda u \qquad (44)$$

for the values of λ (the eigenvalues) for which there exist nonzero solutions u (the eigenfunctions), we assume

$$u = \sum_0^\infty u_n \epsilon^n , \qquad \lambda = \sum_0^\infty \lambda_n \epsilon^n , \qquad (45)$$

where u_0, λ_0 satisfy $L_0 u_0 = \lambda_0 u_0$, and where u_0 is normalized so that $\int u_0^2 = 1$. Our procedure will give that eigenvalue of (44) which approaches λ_0 as $\epsilon \to 0$. We substitute (45) into (44) to obtain first

$$(L_0 + \epsilon L_1)\sum_0^\infty u_n \epsilon^n = (\sum \lambda_m \epsilon^m)\sum u_p \epsilon^p ,$$

and then

$$\sum_{n=0}^\infty \epsilon^n L_0 u_n + \sum_1^\infty L_1 u_{n-1} \epsilon^n = \sum_{n=0}^\infty \epsilon^n \sum_{m=0}^n \lambda_m u_{n-m} .$$

Equating coefficients of ϵ^n, we get

$$L_0 u_n + L_1 u_{n-1} = \lambda_0 u_n + \lambda_1 u_{n-1} + \cdots + \lambda_n u_0 , \tag{46}$$

and in particular

$$L_0 u_0 = \lambda_0 u_0 , \tag{47}$$

which was satisfied from the start, and

$$(L_0 - \lambda_0)u_1 = (\lambda_1 - L_1)u_0 . \tag{48}$$

$$(L_0 - \lambda_0)u_2 = (\lambda_1 - L_1)u_1 + \lambda_2 u_0 , \quad \text{etc.} \tag{49}$$

To solve (48) for λ_1 and u_1, we must have some additional conditions because the homogeneous equation $(L_0 - \lambda_0)v = 0$ has a nonzero solution $v = u_0$. A necessary condition for a solution of (48) is that the inhomogeneous term $(\lambda_1 - L_1)u_0$ be orthogonal to u_0. Thus we require

$$\int u_0 (\lambda_1 - L_1)u_0 = \lambda_1 \int u_0^2 - \int u_0 L_1 u_0 = 0,$$

which defines

$$\lambda_1 = \int u_0 L_1 u_0 \Big/ \int u_0^2 = \int u_0 L_1 u_0 . \tag{50}$$

Substituting (50) into (48) and solving for u_1, we obtain

$$u_1 = v_1 + \alpha_1 u_0 , \tag{51}$$

where v_1 is a particular solution, and where α_1 is an arbitrary constant. We normalize the eigenvector u_1 by taking α_1 such that

$$\int u_0 u_1 = \int u_0 (v_1 + \alpha_1 u_0) = 0;$$

thus

$$\alpha_1 = -\int u_0 v_1 . \tag{52}$$

Similarly, (49) has a solution only if

$$\int u_0 [\lambda_1 u_1 - L_1 u_1 + \lambda_2 u_0] = 0 ,$$

which determines

$$\lambda_2 = \int u_0 L_1 u_1 , \tag{53}$$

and thereby determines the solution of (49) in the form

$$u_2 = v_2 + \alpha_2 u_0 . \tag{54}$$

As before, we normalize by taking α_2 such that $\int u_0 u_2 = 0$, and obtain

$$\alpha_2 = -\int u_0 v_2 . \tag{55}$$

The procedure is general and the eigenfunction u we have constructed satisfies

$$\int u_0 u = 1 = \int u_0 (u_0 + \epsilon u_1 + \cdots) = \int u_0^2 + \epsilon \int u_0 u_1 + \epsilon^2 \int u_0 u_2 + \cdots , \tag{56}$$

where $\int u_0^2 = 1$, and the other terms of the right-hand side are zero by our normalization procedure.

8.3 EXAMPLE OF SINGULAR PERTURBATION PROBLEM

If the simple perturbation procedures we have discussed break down, then we say we are dealing with a singular perturbation problem. As an example, we consider a problem given by Rayleigh.

Consider the fourth-order equation

$$\epsilon \frac{d^4 u}{dx^4} - \frac{d^2 u}{dx^2} = \epsilon u^{(IV)} - u'' = \lambda u \tag{57}$$

with the boundary conditions

$$u(0) = u'(0) = u(1) = u'(1) = 0 . \tag{58}$$

We would like to find those values of λ for which there exists a nonzero solution. This problem occurs in the study of the modes of vibration of a rod; the quantity ϵ is related to the stiffness, and for $\epsilon \sim 0$ the rod behaves like an elastic string. If $\epsilon = 0$, the equation reduces to

$$u'' = -\lambda u \tag{59}$$

which is the equation for the vibration of the string. However, we still have the four boundary conditions of (58), and for any λ the only solution of (59) and (58) is identically zero; thus the previous perturbation methods break down before we can even begin. This is a typical singular perturbation problem. Because the order of the differential equation is lowered in going from (57) to (59), we shall find that some of the boundary conditions of (58) must be dropped when we actually consider (59). However, it is (57) plus (58) in which we are interested.

Since (57) has constant coefficients, the solutions are exponential in form. Substituting $u = \epsilon^{rx}$ in (57), we obtain the characteristic equation

$$r^4 \epsilon - r^2 = \lambda ,$$

which we solve for r^2 to get

$$r^2 = \frac{1 \pm \sqrt{1 + 4\lambda\epsilon}}{2\epsilon} .$$

We are interested primarily in the case where $\epsilon = |\epsilon| = \eta^2$ is small. Thus for the positive radical we find

$$r^2 \approx \frac{1 + 1 + 2\lambda\epsilon}{2\epsilon} \sim \frac{1}{\epsilon} = \frac{1}{\eta^2} , \qquad r = \pm r_1 \approx \pm\frac{1}{\eta} \tag{60}$$

and for the negative,

$$r^2 \approx \frac{1 - (1 + 2\lambda\epsilon)}{2\epsilon} \sim -\lambda , \qquad r = \pm i r_2 = \pm i \sqrt{\lambda} . \tag{61}$$

We take as fundamental solutions of (57) the following four functions: $\epsilon^{r_1 x}$, $e^{-r_1 x}$, $\sin r_2 x$, and $\cos r_2 x$.

The general solution of (57) may be written as

$$u = \alpha_+ e^{r_1 x} + \alpha_- e^{-r_1 x} + \beta \sin r_2 x + \gamma \cos r_2 x , \tag{62}$$

where the constants $\alpha_+, \alpha_-, \beta, \gamma$ must be so determined that the boundary conditions (58) on $u(x)$ are satisfied:

$$u(0) = \alpha_+ + \alpha_- + 0 + \gamma = 0$$

$$u'(0) = \alpha_+ r_1 - \alpha_- r_1 + r_2 \beta + 0 = 0$$

$$u(1) = \alpha_+ e^{r_1} + \alpha_- e^{-r_1} + \beta \sin r_2 + \gamma \cos r_2 = 0 \tag{63}$$

$$u'(1) = \alpha_+ r_1 e^{r_1} - \alpha_- r_1 e^{r_1} + \beta r_2 \cos r_2 - \gamma r_2 \sin r_2 = 0 .$$

These equations will have a nontrivial solution for $\alpha_+, \alpha_-, \beta, \gamma$ if and only if the determinant of the coefficients is zero:

$$
\begin{vmatrix}
1 & 1 & 0 & 1 \\
r_1 & -r_1 & r_2 & 0 \\
e^{r_1} & e^{-r_1} & \sin r_2 & \cos r_2 \\
r_1 e^{r_1} & -r_1 e^{-r_1} & r_2 \cos r_2 & -r_2 \sin r_2
\end{vmatrix} = 0 .
$$

We multiply the first column of the determinant by $r_1^{-1} e^{-r_1}$ and the second column by r_1^{-1} to obtain

$$
\begin{vmatrix}
r_1^{-1} e^{-r_1} & r_1^{-1} & 0 & 1 \\
e^{-r_1} & -1 & r_2 & 0 \\
r_1^{-1} & r_1^{-1} e^{-r_1} & \sin r_2 & \cos r_2 \\
1 & -e^{-r_1} & r_2 \cos r_2 & -r_2 \sin r_2
\end{vmatrix} = 0 .
$$

Since $r_1 \sim \eta^{-1}$, we see both e^{-r_1} and r_1^{-1} vanish in the limit $\eta \to 0$, and the determinant becomes

$$
\begin{vmatrix}
0 & 0 & 0 & 1 \\
0 & -1 & r_2 & 0 \\
0 & 0 & \sin r_2 & \cos r_2 \\
1 & 0 & r_2 \cos r_2 & -r_2 \sin r_2
\end{vmatrix} = \sin r_2 = 0 ,
$$

which requires

$$ r_2 = n\pi , \qquad n = 1, 2, 3, \ldots . \tag{64} $$

Since in the first approximation $r_2 \sim \lambda^{\frac{1}{2}}$, we find that

$$ \lambda = (n\pi)^2, \qquad n = 1, 2, 3, \ldots . \tag{65} $$

Using (64) in (63), we find that

$$ \alpha_+ = \alpha_- = \gamma = 0 . $$

Thus the limiting form of the solution of (57) is

$$u = \sin n\pi x \, , \tag{66}$$

which is the solution of (59) subject to only two of the original boundary conditions as in (58), namely $u(0) = u(1) = 0$.

Boundary Layer Effect. To understand what happened to the other boundary conditions, we shall "magnify" the solution of (57) in the neighborhood of $x = 0$ and $x = 1$. Let us consider the neighborhood of $x = 0$, and write $x = \sqrt{\epsilon}\, t = \eta t$; then (57) becomes

$$\ddddot{u} - \ddot{u} = \lambda \eta^2 u \, , \tag{67}$$

where the dot indicates differentiation with respect to t. Note that for an appropriate small value of x, say $x = \epsilon^{\frac{1}{4}} = \eta^{\frac{1}{2}}$, the corresponding value of t, that is, $t = \eta^{-\frac{1}{2}}$, is large; thus, a small neighborhood of $x = 0$ becomes a very large neighborhood of $t = 0$. Since $x = 0$ corresponds to $t = 0$ and since $u = \sin n\pi x$ is the limiting solution of (57) for $\eta \to 0$, we shall solve (67) by imposing the boundary conditions

$$u(0) = \dot{u}(0) = 0 \, , \tag{68}$$

and the condition

$$u \sim \sin n\pi x = \sin n\pi \eta t \, , \quad \text{as } t \sim \infty \, , \tag{69}$$

which also implies $\lambda \sim (n\pi)^2$ as in (65).

The characteristic equation for (67) is

$$r^4 - r^2 = \lambda \eta^2 \tag{70}$$

which we rewrite in the following two ways:

$$r^2 = 1 + \frac{\lambda \eta^2}{r^2} \, , \quad r^2 = \frac{-\lambda \eta^2}{1 - r^2} \approx \frac{-(n\pi\eta)^2}{1 - r^2} \, .$$

From the first equation, $r = \pm 1 + O(\eta^2)$, and from the second, $r \approx \pm i n\pi\eta$; consequently, the solutions of (67) are approximately e^t, e^{-t}, $\sin n\pi\eta t$, $\cos n\pi\eta t$. The combinations

$$u_1 = n\pi\eta(e^t - \cos n\pi\eta t) - \sin n\pi\eta t \, ,$$

$$u_2 = n\pi\eta(e^{-t} - \cos n\pi\eta t) + \sin n\pi\eta t$$

satisfy the conditions $u(0) = \dot{u}(0) = 0$ of (68), and so does the general form

$$u = \alpha u_1 + \beta u_2 \, ,$$

where α and β are to be determined by the condition (69) that u approach $\sin n\pi\eta t$ as t approaches infinity. Since $u_1 \to \infty$ if t does, we see that we must take $\alpha = 0$, $\beta = 1$. Thus we obtain

$$u_2 = \sin n\pi x + n\pi\eta (e^{-x/\eta} - \cos n\pi x) \,, \quad \eta = \sqrt{\epsilon} \tag{71}$$

as a more accurate solution of (57) than (66) near $x = 0$. The sketch for $n = 1$ in Figure 8.1 shows that (71) satisfies all the boundary conditions at zero, but that in a "boundary layer" whose thickness depends on the magnitude of $\sqrt{\epsilon}$ it "loses" a boundary condition and merges with the solution $\sin n\pi x$ of the limiting equation (59). A similar discussion in terms of $1 - x = \eta s$ with $u = du/ds = 0$ at $s = 0$ leads to the analogous results in the neighborhood of $x = 1$ shown in the sketch.

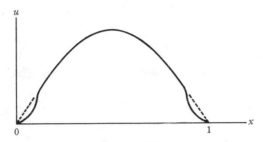

Figure 8.1

Correction to Eigenvalues. We now show how to obtain a correction term to the approximate value of λ given by (65). Consider, again, the neighborhood of $x = 0$ and put $x = \eta t$ as before. Let us use r_1 to denote that root of the characteristic equation (70) which is close to unity, and $r = ir_2$ to denote that root close to $i\eta\sqrt{\lambda}$. The solutions of (67) can be taken as

$$e^{r_1 t}, \quad e^{-r_1 t}, \quad \sin r_2 t, \quad \cos r_2 t,$$

and we require combinations that satisfy (68) and (69). The functions

$$u_1 = (e^{r_1 t} - \cos r_2 t) - \frac{r_1}{r_2} \sin r_2 t \tag{72}$$

$$u_2 = (e^{-r_1 t} - \cos r_2 t) + \frac{r_1}{r_2} \sin r_2 t \tag{73}$$

are solutions of (67) which satisfy the conditions $u(0) = \dot{u}(0) = 0$ of (68). Since $u_1 \to \infty$ as $t \to \infty$, we conclude just as before that the solution of (67) must be proportional to u_2.

Let us now consider the neighborhood of $x = 1$. Put $1 - x = \eta s$, then (57) becomes (67) where now the dots mean differentiation with respect to s. Note that for x close to 1, say $x = 1 - \eta^{\frac{1}{2}}$, the value of s is $\eta^{-\frac{1}{2}}$ is large; at $x = 1$, $s = 0$. Since u and $\dot{u}$ vanish at $s = 0$, we find that the solution of (67) satisfying the conditions at $x = 1$ must be proportional to

$$u_4 = (e^{-r_1 s} - \cos r_2 s) + \frac{r_1}{r_2} \sin r_2 s .$$ (74)

The solutions u_2 and u_4 must match as $t \to \infty$ and $s \to \infty$, that is, for x in the midrange between zero and unity where both are subject to (69). Put

$$\tan \Theta = r_2/r_1 ;$$ (75)

then for large t we may recast (73) as

$$u_2 = \sin\left(\frac{r_2 x}{\eta} - \Theta\right),$$ (76)

and take the corresponding form of (74) for large s to be

$$u_4 = \sin\left[r_2 \frac{(1 - x)}{\eta} - \Theta\right].$$ (77)

Since these functions are proportional for all x not too close to zero and unity we have

$$\frac{u'_2}{u_2} = \frac{u'_4}{u_4}$$

or

$$\cot\left(\frac{r_2}{\eta} x - \Theta\right) = - \cot\left[r_2 \frac{(1 - x)}{\eta} - \Theta\right].$$

This relation will be satisfied if

$$\frac{r_2 x}{\eta} - \Theta + \frac{r_2(1 - x)}{\eta} - \Theta = n\pi ,$$ (78)

where n is any positive or negative integer. From (78) we find

$$r_2 = \eta n\pi + 2\eta\Theta .$$ (79)

Since from (70), $r^2 = (ir_2)^2 \approx -\lambda\eta^2(1 - \lambda\eta^2)$, we have

$$r_2 = \eta\lambda^{\frac{1}{2}} + O(\lambda^{\frac{3}{2}} \eta^3),$$

and since from (75),

$$\Theta \sim \frac{r_2}{r_1} \sim \eta \lambda^{\frac{1}{2}}$$

for small η, (79) becomes

$$\lambda^{\frac{1}{2}} \approx n\pi + 2\eta \lambda^{\frac{1}{2}} .$$

Thus

$$\lambda^{\frac{1}{2}} \approx n\pi(1 + 2\eta)$$

and

$$\lambda \approx n^2\pi^2(1 + 4\eta) \tag{80}$$

replaces (65). We have thus obtained the first correction to the eigenvalues of the unperturbed problem (59). The corresponding correction of the unperturbed eigenfunctions $\sin n\pi x$ may be obtained from (76):

$$u \approx \sin[\sqrt{\lambda}(x - \eta)] \approx \sin n\pi x - n\pi\eta \cos n\pi x . \tag{81}$$

8.4 SINGULAR PERTURBATION PROCEDURE

We now consider differential equations of the general form

$$\epsilon[p_0(x)u^{(VI)} + p_1(x)u^{(V)} + \cdots + p_6(x)u] + [q(x)u'' + q_1(x)u' + q_2(x)u] = f(x) \tag{82}$$

for $a \le x \le b$, with the boundary conditions

$$B_j(u) = \alpha_{0j}u(a_j) + \alpha_{1j}u'(a_j) + \cdots + \alpha_{5j}u^{(V)}(a_j) = 0, \quad 1 \le j \le 6, \tag{83}$$

where $a_j = a$ or b, the end points of the interval. As $\epsilon \to 0$ the differential equation becomes second order, but we have six boundary conditions, and as before this causes difficulty.

We represent equation (82) by

$$[\epsilon L + M]u = f , \tag{84}$$

and introduce g as a particular solution satisfying

$$[\epsilon L + M]g = f . \tag{85}$$

Then we may write the general solution as

$$u = g + V , \quad (\epsilon L + M)V = 0 , \tag{86}$$

where V satisfies the inhomogeneous equation, and also the boundary conditions

$$B_j(V) = -B_j(g), \qquad 1 \le j \le 6.$$ (87)

We expand V in powers of ϵ,

$$V = \sum_{n=0}^{\infty} V_n(x)\epsilon^n$$ (88)

and substitute in (84) to obtain

$$LV_n + MV_{n+1} = 0,$$

$$MV_0 = q_0 V_0'' + q_1 V_0' + q_2 V_0 = 0,$$ (89)

$$MV_1 = -LV_0 = -[p_0 V_0^{(VI)} + \cdots + p_6 V_0],$$

etc. Thus, knowing V_0, we can find successive terms of the expansion (88). In general, the series will not converge but will be asymptotic to the solution.

To analyze the expansion for V_1 we must know how the solutions of (82) behave as $\epsilon \to 0$. It can be shown that as $\epsilon \to 0$ two linearly independent solutions of (84) converge to two independent solutions of $Mu = 0$. Let us denote these solutions by $y_1(x)$ and $y_2(x)$. What happens to the other four linearly independent solutions? To discover this, let us consider the special case where (82) reduces to

$$\epsilon u^{(VI)} + u'' = 0.$$ (90)

The solutions of this equation are of the form e^{rx}, where r satisfies the characteristic equation

$$\epsilon r^6 + r^2 = r^2(\epsilon r^4 + 1) = 0.$$ (91)

The roots of this equation are $r = 0, 0$, and r equal to the four fourth-roots of $-\epsilon^{-1}$. In terms of $\epsilon = -\eta^4$, we write the roots as

$$r = 0, \quad 0, \quad \pm\eta^{-1}, \quad \pm i\eta^{-1},$$ (92)

and take the linearly independent solutions of (90) as

$$u = 1, \quad x, \quad e^{\eta^{-1}x}, \quad e^{-\eta^{-1}x}, \quad e^{i\eta^{-1}x}, \quad e^{-i\eta^{-1}x}.$$ (93)

As $\epsilon \to 0$, or equivalently $\eta \to 0$, the first two of these functions become the solution of the limiting equation

$$u'' = 0;$$ (94)

the others either converge to zero or do not converge at all.

This example suggests that the solutions of (82) which do not converge to a solution of $MV = 0$ may contain exponential factors. We shall show that this is correct. We put $\epsilon = -\eta^4$, and seek a solution of (82) in the form

$$u = e^{\eta^{-1}k(x)}\sum_0^\infty W_n(x)\eta^n = e^{\eta^{-1}k(x)}t(x, \eta) . \tag{95}$$

We have

$$u' = \eta^{-1} k'u + e^{\eta^{-1}k}t' = [\eta^{-1}k' + D]u ,$$

where D will indicate differentiation of the factor multiplying the exponential. Continuing the differentiation, we get

$$u'' = [\eta^{-1}k' + D]^2 u = (\eta^{-1}k' + D)(\eta^{-1}k' + D)u$$

$$= \eta^{-2}k'^2 + \eta^{-1}k'D + D[\eta^{-1}k'u] + D^2u .$$

But

$$D[\eta^{-1}k'u] = \eta^{-1}k''u + \eta^{-1}k'Du ;$$

consequently

$$u'' = (\eta^{-2}k'^2 + 2\eta^{-1}k'D + D^2 + \eta^{-1}k'')u .$$

We guess that

$$(\eta^{-1}k' + D)^n u \sim \eta^{-n}(k')^n + \binom{n}{1}\eta^{-n+1}(k')^{n-1} D \tag{96}$$

$$+ \binom{n}{2}\eta^{-n+1}(k')^{n-2} k'' + O(\eta^{-n+2}) .$$

This is correct for $n = 1, 2$ and can be proved by induction for all integral values of n. Substituting (95) into (82) and using (96), we obtain

$$-\eta^4[p_0\{\eta^{-6}k'^6 + 6\eta^{-5}k'^5 D + 15\eta^{-5}k'^4 k''\}$$

$$+ p_1\{\eta^{-5} k'^5 + 5\eta^{-4}k'^4 D + 10\eta^{-4}k'^3 k''\} + p_2\{\eta^{-4}k'^4\} + O(\eta^{-4})]u$$

$$+ [q_0\{\eta^{-2}k'^2 + 2\eta^{-1} k'D + D^2 + \eta^{-1}k''\} + q_1(\eta^{-1}k' + D) + q_2]u = 0 .$$

Comparing the coefficients of corresponding powers of η, we get

$$-p_0k'^6 + q_0k'^2 = 0 \tag{97}$$

and

$$(-6k'^5 p_0 + 2k'q_0)W_0' + \{-15k'^4 k''p_0 + k'^5p_1 + k''q_0 + k'q_1\}W_0 = 0 . \tag{98}$$

Equation (97) has two solutions with k' equal to zero. These two solutions correspond to the linearly independent solutions

$$y_1(x) , \qquad y_2(x) \tag{99}$$

which converge as $\epsilon \to 0$ to the solutions of $Mu = 0$. Let $r(x)$ be any nonzero solution of (97), that is, a solution of

$$-p_0 r^4 + q_0 = 0 . \tag{100}$$

Then

$$k(x) = \int_c^x r(\xi)\, d\xi$$

will satisfy (97). Knowing $k(x)$, we can solve (98) for W_0 if the coefficient of W_0' is not zero; since the coefficient $-6p_0 r^5 + 2q_0 r$ is not zero if $r \neq 0$, we see that W_0 can be determined from (98). Using the coefficients of the other powers of η, we may successively determine the functions $W_1, W_2, \ldots$, etc., thus obtaining the expansion (95). Since there are four distinct roots of (100), this procedure will give the expansion of four linearly independent solutions of (82); we will denote these by

$$C_j(x) = t_j(x, \eta) \exp[\eta^{-1} \int_c^x r_j(x)] = t_j \exp[\eta^{-1} k_j(x)] , \quad 1 \le j \le 4, \tag{101}$$

where $r_j(x)$ is a solution of (100) such that r_1 and r_3 have positive real parts, and $r_2 = -r_1, r_4 = -r_3$.

Note that in general, the expansion (95) we have constructed will not converge, but it can be shown that it is asymptotic to a solution of (82).

Let us return to the consideration of (84) with the boundary conditions (83). The general solution of (84) may be written in terms of the y's and C's introduced in (99) and (101) as

$$u = g + \beta_1 y_1 + \beta_2 y_2 + \gamma_1 C_1 + \gamma_2 C_2 + \delta_1 C_3 + \delta_2 C_4 \tag{102}$$

where the values of $\beta_1, \beta_2, \gamma_1, \gamma_2, \delta_1, \delta_2$ must be determined to fit the boundary conditions (83). In terms of (102), these boundary conditions become

$$B_j(u) = B_j(g) + \beta_1 B_j(y_1) + \cdots + \delta_2 B_j(C_4) = 0, \quad 1 \le j \le 6 . \tag{103}$$

These six linear equations for the six unknowns can be solved by determinants. Consider the determinant of the coefficients, namely,

$$
\Delta = \begin{vmatrix}
B_1(y_1) & B_1(y_2) & B_1(C_1) & B_1(C_2) & B_1(C_3) & B_1(C_4) \\
B_2(y_1) & B_2(y_2) & B_2(C_1) & B_2(C_2) & B_2(C_3) & B_2(C_4) \\
B_3(y_1) & B_3(y_2) & B_3(C_1) & B_3(C_2) & B_3(C_3) & B_3(C_4) \\
B_4(y_1) & B_4(y_2) & B_4(C_1) & B_4(C_2) & B_4(C_3) & B_4(C_4) \\
B_5(y_1) & B_5(y_2) & B_5(C_1) & B_5(C_2) & B_5(C_3) & B_5(C_4) \\
B_6(y_1) & B_6(y_2) & B_6(C_1) & B_6(C_2) & B_6(C_3) & B_6(C_4)
\end{vmatrix}
$$

Note that the C_k and their derivatives have an exponential factor which goes to either zero or infinity as η goes to zero. This will depend on whether the real part of the exponent is positive or negative. However, with some manipulation we can ensure that the terms with the exponentials will go to zero; the procedure is as follows:

Multiply the third column of Δ by $\exp[-\eta^{-1}k_1(b)]$, the fourth column by $\exp[\eta^{-1}k_1(a)]$, the fifth column by $\exp[-\eta^{-1}k_3(b)]$, and the sixth by $\exp[\eta^{-1}k_3(a)]$. The jth term in the third column will therefore have as an exponential factor either $\exp[\eta^{-1}\{k_1(a) - k_1(b)\}]$ or $\exp[\eta^{-1}\{k_1(b) - k_1(b)\}]$, depending on whether B_j is a boundary condition at $x = a$ or $x = b$. Since

$$
k_1(b) - k_1(a) = \int_a^b r_1(\xi)\,d\xi
$$

and since the real part of $r_1(x)$ is positive, it follows that the real part of $k_1(b) - k_1(a)$ is positive; consequently,

$$
\exp[\eta^{-1}\{k_1(a) - k_1(b)\}]
$$

goes to zero as η goes to zero, and

$$
\exp[\eta^{-1}\{k_1(b) - k_1(b)\}] = 1
$$

is bounded. This means that those $B_j(C_1)$ which are to be evaluated at $x = a$ are to be omitted in the limit. Similarly, those terms of $B_j(C_2)$

which are to be evaluated at $x = b$ are to be omitted. A similar argument
works for $B_j(C_3)$ at $x = a$, and for $B_j(C_4)$ at $x = b$.

A further simplification can be made. Consider the terms in the third
column which are to be evaluated at $x = b$. Even though the exponential
factors are bounded, there are still inverse powers of η involved because
of the derivatives of C_1 in $B_j(C_1)$. We combine these boundary conditions
in such a way that no two of them have the same highest-order derivative
and then rewrite them so that the highest-order derivative in B_j is of
lower order than the highest-order derivative in B_k if $j < k$. Let p be
the highest-order derivative in the $Bj(C_1)$ which are to be evaluated
at $x = b$. Multiply the third column by η^p and let η go to zero. Then
only one term will remain in the third column. A similar argument will
show that only one term remains in each of the fourth, fifth, and sixth
columns.

Let us illustrate this argument by considering the special case in
which the boundary conditions are

$$u(a) = 0, \qquad\qquad u(b) + u'(b) = 0,$$

$$u'(a) + u''(a) = 0, \quad u'(b) + u'''(b) = 0,$$

$$u(a) + u''(a) = 0, \quad u''(b) + u'''(b) = 0.$$

We see that the boundary conditions at $x = a$ are equivalent to

$$u(a) = u'(a) = u''(a) = 0,$$

and the boundary conditions at $x = b$ are equivalent to

$$u'''(b) + u''(b) = u''(b) - u'(b) = u'(b) + u(b) = 0.$$

In this form the highest-order derivative in each boundary condition does
not appear as a highest-order derivative in any other boundary condition.

The leading term in the determinant Δ in this case is

$$\Delta \sim \begin{vmatrix} B_1(y_1) & B_1(y_2) & e_1(a) & e_1(a)^{-1} & e_3(a) & e_3(a)^{-1} \\ B_2(y_1) & B_2(y_2) & \eta^{-1}e_1(a) & \eta^{-1}e_1(a)^{-1} & \eta^{-1}e_3(a) & \eta^{-1}e_3(a)^{-1} \\ B_3(y_1) & B_3(y_2) & \eta^{-2}e_1(a) & \eta^{-2}e_1(a)^{-1} & \eta^{-2}e_3(a) & \eta^{-2}e_3(a)^{-1} \\ B_4(y_1) & B_4(y_2) & \eta^{-3}e_1(b) & \eta^{-3}e_1(b)^{-1} & \eta^{-3}e_3(b) & \eta^{-3}e_3(b)^{-1} \\ B_5(y_1) & B_5(y_2) & \eta^{-2}e_1(b) & \eta^{-2}e_1(b)^{-1} & \eta^{-2}e_3(b) & \eta^{-2}e_3(b)^{-1} \\ B_6(y_1) & B_6(y_2) & \eta^{-1}e_1(b) & \eta^{-1}e_1(b)^{-1} & \eta^{-1}e_3(b) & \eta^{-1}e_3(b)^{-1} \end{vmatrix},$$

where

$$e_1(x) = e^{\eta^{-1} k_1(x)}, \qquad e_3(x) = e^{\eta^{-1} k_3(x)} .$$

Multiply the third column by $\eta^3 e_1(b)^{-1}$, the fourth column by $\eta^2 e_1(a)$, the fifth column by $\eta^3 e_3(b)^{-1}$, and the sixth column by $\eta^2 e_3(a)$. Using the fact that $e_1(a) e_1(b)^{-1}$ and $e_3(a) e_3(b)^{-1}$ go to zero as η goes to zero, we see that

$$\eta^{10} e_1(a) e_1(b)^{-1} e_3(a) e_3(b)^{-1} \Delta = \begin{vmatrix} B_1(y_1) & B_1(y_2) & 0 & \eta^2 & 0 & \eta^2 \\ B_2(y_1) & B_2(y_2) & 0 & \eta & 0 & \eta \\ B_3(y_1) & B_3(y_2) & 0 & 1 & 0 & 1 \\ B_4(y_1) & B_4(y_2) & 1 & 0 & 1 & 0 \\ B_5(y_1) & B_5(y_2) & \eta & 0 & \eta & 0 \\ B_6(y_1) & B_6(y_2) & \eta^2 & 0 & \eta^2 & 0 \end{vmatrix} + O(\eta^3)$$

$$= A\eta^2 \begin{vmatrix} B_1(y_1) & B_1(y_2) \\ B_6(y_1) & B_6(y_2) \end{vmatrix} + O(\eta^3) ,$$

where A is a constant independent of η. This shows that only the first and last boundary conditions, namely,

$$u(a) = 0 , \qquad u'(b) + u(b) = 0 ,$$

are significant.

A similar argument can be used on the determinants in the numerator of the ratios which are the solutions of the linear equations (103). If this is done, we find that $\gamma_1, \gamma_2, \delta_1, \delta_2$ go to zero as η goes to zero; consequently, the limiting problem will reduce to the differential equation

$$Mu = f$$

with the boundary conditions

$$u(a) = 0 , \qquad u'(b) + u(b) = 0 .$$

BIBLIOGRAPHY

Boole, G., *Calculus of Finite Differences*, 1860 ed., reprinted by Chelsea, New York.

Boole, G., *Differential Equations*, 1865 ed., reprinted by Chelsea, New York.

Brand, Louis, *Differential and Difference Equations*, Wiley, New York, 1966.

Bromwich, T. J. I'a., *Theory of Infinite Series*, Macmillan, London, 1949.

Carrier, G. C., M. Krook, and C. E. Pearson, *Functions of a Complex Variable; Theory and Techniques*, McGraw-Hill, New York, 1960.

Copson, E. T., *Asymptotic Expansions*, Cambridge University Press, New York, 1965.

de Bruijn, N. G., *Asymptotic Methods in Analysis*, North-Holland, Amsterdam, 1961.

Erdélyi, A., *Asymptotic Expansions*, Dover, New York, 1956.

Erdélyi, A., et al., *Higher Transcendental Functions*, McGraw-Hill, New York, 1953.

Friedman, B., *Principles and Techniques of Applied Mathematics*, Wiley, New York, 1956.

Hadamard, J., *Lectures on Cauchy's Problem in Linear Partial Differential Equations*, pp. 133ff, Dover, New York, 1952.

Parzen, Emanuel, *Modern Probability Theory and its Applications*, Wiley, New York, 1960.

Schwartz, L., *Mathematics for the Physical Sciences*, Addison-Wesley, Reading, Mass., 1966.

Schwartz, L., *Theorie des Distributions*, Vol. 1 and 2, Actualités Scientifique et Industrielles, Hermann & Cie, Paris, 1950 and 1951.